THE LIVES OF
SHARKS

THE LIVES OF SHARKS

A NATURAL HISTORY OF SHARK LIFE

Daniel C. Abel
& R. Dean Grubbs

PRINCETON UNIVERSITY PRESS
PRINCETON AND OXFORD

Published by Princeton University Press
41 William Street, Princeton, New Jersey 08540
99 Banbury Road, Oxford OX2 6JX
press.princeton.edu

Copyright © 2023 by UniPress Books Limited
www.unipressbooks.com

All rights reserved. No part of this book may be reproduced or transmitted in any form or by any means, electronic or mechanical, including photocopying, recording, or by any information storage-and-retrieval system, without written permission from the copyright holder. Requests for permission to reproduce material from this work should be sent to permissions@press.princeton.edu

Library of Congress Control Number 2022947655
ISBN 978-0-691-24431-0
Ebook ISBN 978-0-691-25220-9

Typeset in Bembo and Futura

Printed and bound in Latvia
10 9 8 7 6 5 4 3 2 1

British Library Cataloging-in-Publication Data is available

This book was conceived, designed, and produced by
UniPress Books Limited
Publisher: Nigel Browning
Commissioning editor: Kate Shanahan
Project manager: Richard Webb
Designer & art direction: Wayne Blades
Picture researcher: Tom Broadbent
Illustrator: Sarah Skeate
Maps: Les Hunt

Cover images: (Front cover) Cultura Creative Ltd/Alamy Stock Photo; (back cover and spine) Kletr/Shutterstock

Cover design: Wayne Blades and Wanda España

CONTENTS

6
INTRODUCTION

18
THE EVOLUTION AND
DIVERSITY OF SHARKS

48
ADAPTATIONS OF SHARKS

84
THE ECOLOGY OF SHARKS

116
SHARKS OF THE OPEN OCEAN

146
SHARKS OF THE DEEP SEA

174
SHARKS OF ESTUARIES
AND RIVERS

206
SHARKS OF THE
CONTINENTAL SHELVES

240
SHARKS AND US

276 Glossary
280 Resources
284 Index
288 Acknowledgments

INTRODUCTION

INTRODUCTION

The world of sharks

Every May for nearly 30 years we have taught a course in shark biology at the Bimini Biological Field Station in The Bahamas. Within hours of our arrival in this tropical paradise, we repeat a ritual that culminates in what our students have been waiting their entire lives to experience: snorkeling with wild, actively feeding, 8 ft (2.5 m) Caribbean Reef Sharks (*Carcharhinus perezi*; page 234).

When you close your eyes and envision a shark, the chances are it looks like a Caribbean Reef Shark: a graceful, magnificent, sleek, heavily muscled gray predator. These are the beasts of your dreams if you're a shark lover, or perhaps your nightmares if you aren't. Surprisingly, the students are silent in the skiffs as they await our short boat ride, and the air is tinged with a palpable dose of nervous anticipation mixed with a sense of impending terror. You can almost read some of their minds as they question the wisdom of enrolling in the course. *What have we gotten ourselves into?*

We motor 15 minutes over almost brazenly azure seas to our destination, a stony outcrop known as Triangle Rocks. Within minutes, as the students prepare to slip into the crystal-clear water, large Caribbean Reef Sharks appear one by one and execute a series of ever-tightening circles around our boats, drawn from perhaps a mile or more away by the sound of our outboard motors. This sound informs them that the aquatic equivalent of a buffet might soon commence, and who in their right mind passes up free food?

The sharks have done this before, so they know the drill: wait until the students are aligned shoulder to shoulder, grasping a floating rope stretching from the stern of the biggest boat, then the feeding will start, as chunks of fish are tossed into the water. While the sharks may know what awaits, the students do not, since this is a novel experience for them. It's one thing to watch divers observing sharks from the safety of a metal cage on TV, but it's a totally foreign experience to be immersed into a mob of 8 ft sharks that suddenly look much, much larger than the students envisioned.

What brought the students and us to this point is an unabashed fascination with, and love of, sharks—animals we have been conditioned our entire lives to fear. But we, our students, and you, we suspect, rise above that fear. Instead, we marvel at the prowess of sharks as predators and their undeniable grace, even as they devour their unlucky prey. We are mesmerized by their presence, whether on TV, in an aquarium, or especially, in nature. We want to know more about them. What did sharks look like early in their evolutionary history? How many kinds of sharks are there and how do they differ from one another? Do sharks live in rivers and lakes? What makes them such effective predators? Are we justified in fearing them? How stable are shark populations and what threatens them? What would the oceans be like without sharks? What can we do to make sure that doesn't happen?

→ Caribbean Reef Sharks (*Carcharhinus perezi*) silhouetted against downwelling light under a dive boat in the clear tropical waters of The Bahamas.

Surprisingly, within minutes of arriving at Triangle Rocks, almost all the students are in position to watch what the public might call a feeding frenzy but which is more aptly described as "the dance of the sharks." Some students jockey to be the first to leave the perceived safety of the boat to swim with the sharks, the number of which by that point has grown to perhaps a dozen. Others are more hesitant, but all eventually transition into the briny sea and take their place shoulder to shoulder on the floating rope. The scene has now transitioned from so-called "shark-infested waters" to what might more appropriately be labeled as "human-infested."

It took 450 million years of evolution to perfect the form, senses, and behavior of these sharks, but it is a mistake to consider them primitive, since other vertebrates—specifically the bony fishes (for example, cod, tuna, and bass) and reptiles—trace their early evolution to the same geological era as sharks. The Caribbean Reef Shark does not look remotely like its early ancestors, although it does possess many of the same characteristics that define it as a shark (pages 22–27). Over their relatively long evolutionary history, the majority of these ancient relatives became extinct, leaving behind what we know as modern sharks. But the Caribbean Reef Shark has one major characteristic in common with both its ancestors and other modern sharks: it is a predator.

When the first chunk of bait hits the water, the dance begins. No movement is wasted or haphazard. The sharks' senses allow them to hear the splash of the bait and to determine its location. The bait's rapidly spreading odor provides more clues to its location, and draws in even more sharks as it spreads into the surrounding area. Watching the last part of the sharks' approach to the chunk of fish that was tossed into the water only seconds earlier is the most fascinating aspect of this spectacle. As one or more sharks move into position to be the first to gobble the bait, their vision

↖ A solo Caribbean Reef Shark cruising the shallow waters of The Bahamas. This species is the most common shark of Caribbean coral reefs and it occupies a similar ecological niche to the Indo-Pacific Grey Reef Shark (Carcharhinus amblyrhynchos).

↑ Lemon Sharks (Negaprion brevirostris), distinguished by their muted yellowish tint and two similarly sized dorsal fins, are found in the shallows of temperate and tropical Atlantic and eastern tropical Pacific coasts.

↓ A diver feeding a Great Hammerhead (Sphyrna mokarran) in Bimini, The Bahamas, while Nurse Sharks (Ginglymostoma cirratum) and a variety of bony fishes scavenge for the scraps.

in this clear water allows them to move swiftly directly toward it, aided by their powerful swimming musculature and ability to fine-tune their movement using their complement of fins. Since their prey in natural circumstances might fight back, the sharks are programmed to close their eyes as they open their mouths about 3 ft (1 m) or so from the bait. If the bait were a live fish, the sharks would detect the minute electrical currents that all living organisms emit, and this signal would guide them to the prey with surgical precision, which was first demonstrated in cat sharks.

After an hour or so of observing the shark ballet, the students are hooked for life. The conversations we overhear the rest of the day reflect the passion that shark lovers all over the world share. Their nervous jitters have been replaced with goose bumps as they reflect on the experience, which most call the greatest of their lives. And it is only the first day of the course! Over the next six days, they will swim with, observe, or

INTRODUCTION

otherwise study Tiger Sharks (*Galeocerdo cuvier*; page 232), Nurse Sharks (*Ginglymostoma cirratum*; page 40), and other species.

Students in the shark biology class will also learn the same basic content we have included in *The Lives of Sharks*. This book delves into sharks in all of their multidimensional facets, from the roles they play in our lives, to how human activities threaten theirs. We discuss how sharks have changed and how they have stayed the same over the course of their 450 million-year evolutionary history. We cover the diversity of shapes, sizes, lifestyles, habitats, and ecological roles of the 500-plus species living today, challenging what you think of as a typical shark. Throughout the book we reflect on the suite of adaptations that make sharks such effective predators and that have allowed them to occupy almost all divisions of the ocean and even some freshwater habitats. We consider the unimaginable situation of what the world's oceans would be like without sharks.

The sharks we have chosen represent a broad cross section of iconic species that we are certain you know, including the White Shark (*Carcharodon carcharias*; page 106) and Whale Shark (*Rhincodon typus*; page 144). But we also include an array of species most readers might not be as familiar with, such as the Goblin Shark (*Mitsukurina owstoni*; page 166) and Swell Shark (*Cephaloscyllium ventriosum*; page 42). And we include one of the most endangered fish in the sea, the Smalltooth Sawfish (*Pristis pectinata*; page 112), which is actually a ray, a shark relative whose body shape and position in the food chain belie the fact that it is not a shark. Not all of our profiled species are apex predators, since most species of shark are not at the very top of the food chain.

Our journey into the lives of sharks is a celebration of these magnificent and endlessly fascinating creatures, but it is also an examination of their biology, ecology, behavior, and conservation. While this book cannot repeat the experience of swimming with sharks, we hope that our love of sharks and our passion for shark conservation shine sufficiently in this book to satisfy your own love of these sometimes misunderstood, infinitely interesting, mysterious predators.

Daniel C. Abel and R. Dean Grubbs

INTRODUCTION

Shark classification

Currently, 36 families and about 543 species of sharks are known to science. Rightfully, since they are all chondrichthyans, we should add the batoids and chimaeras to this list, bringing the total to 61 families and about 1,300 species. Using the classic taxonomic hierarchy of life, all chondrichthyans are in the domain Eukarya, kingdom Animalia, phylum Chordata, subphylum Vertebrata (or Craniata), and superclass Gnathostomata.

Your dog, cat, pet fish, and you yourself are in these same categories. So are all of the bony fishes (minus the relatively few jawless forms), amphibians, reptiles, birds (or avian reptiles), and mammals. Sharks, batoids, and chimaeras are all members of the class Chondrichthyes. Within the Chondrichthyes are two subclasses: Holocephali (chimaeras, or ghost sharks) and Elasmobranchii, the most species-rich and diverse chondrichthyans. Elasmobranchs are further subdivided into three superorders: Batoidea (skates and rays), Squalomorphii (about 179 extant, mostly cold-water, species), and Galeomorphii (about 364 species representing a diverse array of sharks).

A LIST OF SHARK FAMILIES

The 36 taxonomic families of sharks that are currently recognized, along with their common names and the number of species. There are 11 and 25 families in the superorders Squalomorphii and Galeomorphii, respectively.

SUPERORDER

SQUALOMORPHII (DOGFISH SHARKS)

Chlamydoselachidae
Frilled sharks (Two species)

Hexanchidae →
Cow sharks (Five species)

SHARK CLASSIFICATION

Echinorhinidae
Bramble or prickly sharks
(Two species)

Squatinidae
Angel sharks (22 species)

Pristiophoridae
Sawsharks (10 species)

Squalidae
Dogfish sharks (39 species)

Centrophoridae
Gulper sharks (16 species)

Etmopteridae
Lantern sharks (51 species)

Somniosidae
Sleeper sharks (17 species)

Oxynotidae
Rough sharks (Five species)

Dalatiidae
Kitefin sharks (10 species)

SUPERORDER

**GALEOMORPHII
(GALEOMORPH SHARKS)**

Heterodontidae
Bullhead or horn sharks
(Nine species)

Parascylliidae
Collared carpet sharks
(Eight species)

Brachaeluridae
Blind sharks (Two species)

Orectolobidae
Wobbegongs (12 species)

Hemiscylliidae ↑
Bamboo or longtailed
sharks (17 species)

Ginglymostomatidae
Nurse sharks (Four species)

Stegostomatidae ↓
Zebra Shark (One species)

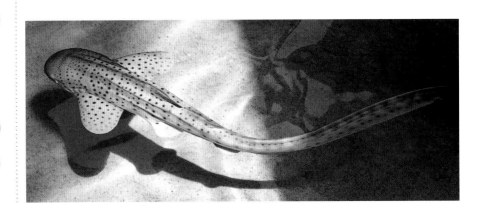

INTRODUCTION

Rhincodontidae ↑
Whale Shark (One species)

Odontaspididae
Sand tigers (Three species)

Pseudocarchariidae
Crocodile Shark (One species)

Mitsukurinidae
Goblin Shark (One species)

Megachasmidae
Megamouth Shark (One species)

Alopiidae
Thresher sharks (Three species)

Cetorhinidae
Basking Shark (One species)

Lamnidae
Mackerel sharks (Five species)

Scyliorhinidae →
Cat sharks (49 species)

Proscylliidae
Finback cat sharks
(Seven species)

Pseudotriakidae
False cat sharks (Six species)

Leptochariidae
Barbeled Houndshark
(One species)

Pentanchidae
Deep-sea cat sharks
(110 species)

Triakidae
Houndsharks (47 species)

Hemigaleidae
Weasel sharks
(Eight species)

Carcharhinidae
Requiem sharks
(57 species)

Galeocerdidae
Tiger Shark (One species)

Sphyrnidae →
Hammerhead sharks
(Nine species)

THE EVOLUTION & DIVERSITY OF SHARKS

THE EVOLUTION AND DIVERSITY OF SHARKS

Sharks in our lives

Our lives intersect with sharks in many ways, as objects of our admiration, fascination, recreation, respect, scholarly study, economy, and unfortunately, as paragons of fear. It would be disingenuous to ignore the last of these, since the primal dread that many harbor for this group of predators overwhelms the positive feelings of others. But any fear we feel should be placed firmly in perspective.

→ The Blacktip Shark (*Carcharhinus limbatus*), a swift-swimming consumer of small schooling fishes, is imputed to be the source of infrequent bite-and-release interactions with swimmers along the southeast US coast.

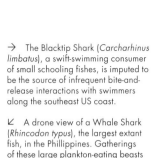

↙ A drone view of a Whale Shark (*Rhincodon typus*), the largest extant fish, in the Phillippines. Gatherings of these large plankton-eating beasts here and in other areas have led to massive influxes of tourists, often to the detriment of the sharks.

As we acknowledge that sharks occasionally bite and, sadly, even kill people, let us equally recognize that these events are extremely rare. Should you be afraid of swimming at the beach? Of course! But not because of sharks. More immediate dangers include being struck in the head by a Jet Ski or surfboard, being exposed to toxic pollutants or unhealthy bacteria, or being sucked into the surf by an undertow or whisked offshore by rip currents.

Both of the authors have been bitten by sharks—numerous times in fact—as have most other shark scientists. In virtually all cases, these bites occur as we handle a shark by removing it from its habitat as part of our research. Consider this: divers, swimmers, and scientists spend countless hours in shark habitat, knowingly or not, with very few negative interactions. But shark bites (called "bites" because they are considered relatively minor, mistaken-identity events) and shark attacks (which imply defensive behavior or involve more serious, repeated contacts) do happen.

The reality is that sharks are far more important to us than as mere objects of our fear. Sharks that are top-level predators stabilize ecosystems, and shark ecotourism bolsters local economies and is a driving force for conservation of marine ecosystems. Sharks are useful in a number of biomedical applications (for example, shark-liver oil is an ingredient in a widely used antihemorrhoidal cream and antibacterial coatings), and they are fished for meat and fins, in some cases sustainably. But paramount to those of us who can transcend the fear factor, sharks captivate us by their beauty, grace, behaviors, and adaptations.

RARE OCCURRENCES

According to the International Shark Attack File, an authoritative database of both shark bites and attacks, in 2021 there were only 79 unprovoked human–shark interactions and 39 provoked bites. These are certainly underestimates, since interactions occurring in more remote or sparsely populated areas are very likely underreported, but even so the true total will still be low.

THE EVOLUTION AND DIVERSITY OF SHARKS

What is a shark?

Unless you are thinking of a Gulper Shark (*Centrophorus granulosus*) or Swell Shark (*Cephaloscyllium ventriosum*; page 42), when you envision a shark you are likely thinking of a streamlined, large (more than 3 ft/1 m) gray shark that lives along shorelines, such as a Blacktip Reef Shark (*Carcharhinus melanopterus*) or Great White Shark (*Carcharodon carcharias*; page 106), which is referred to from now on by its accepted common name, White Shark. The iconic species you conjure up seem representative of sharks as a group, since these are the species we interact with, mostly through the media but in many cases when we dive with them. However, these are, in fact, the oddballs.

To see a typical shark you must either visit the deep sea (defined as ocean deeper than 650 ft/200 m), or be a deep-sea commercial fisher or shark biologist. That is because about two-thirds of the 543 known species of living sharks are 3 ft (1 m) or smaller in length and more than half of them live in the deep sea. Perhaps to the dismay of those who watch sensationalized shark documentaries on TV, a typical shark is small and brown, and lives in the deep ocean.

The new normal

You might consider large coastal sharks like the Blacktip Reef Shark (*Carcharhinus melanopterus*) or Bull Shark (*Carcharhinus leucas*) as typical, but small, deep-sea sharks, such as the Gulper and Cookiecutter Sharks (*Isistius brasiliensis*), are in fact more representative.

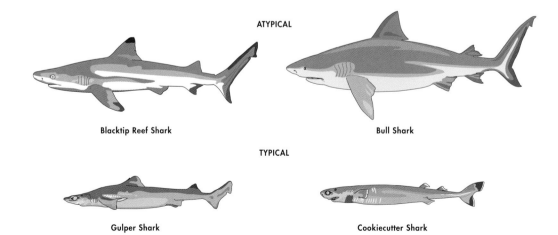

ATYPICAL

Blacktip Reef Shark

Bull Shark

TYPICAL

Gulper Shark

Cookiecutter Shark

The environmental conditions of the deep sea are very different from those of shallow ecosystems, to the extent that only a handful of sharks are able to live in or travel between both. Living in the cold, dark, food-poor deep-sea environment means evolving adaptations to conserve energy, locate prey, avoid being prey yourself, and find mates. And these adaptations are the sources of departures from the fast-swimming, large-bodied gray forms of most coastal sharks. However, there is one way in which deep-sea sharks are typical: in several cases their populations are threatened by human impacts.

TOP PREDATORS

Contrary to popular belief, not all sharks are apex predators—that is, they are not at the top of the food chain. However, all sharks—even plankton-eating species such as Whale and Basking Sharks (*Rhincodon typus* and *Cetorhinus maximus*, respectively)—are predators. Most shark species, including the Port Jackson Shark (*Heterodontus portusjacksoni*) and Nurse Shark (*Ginglymostoma cirratum*), plus all of the dogfish sharks and smooth-hound sharks, would be considered mesopredators, one or more levels below the top shark predators, which include the White Shark, Bull Shark (*Carcharhinus leucas*), and Oceanic Whitetip Shark (*C. longimanus*).

Predators often play disproportionately large roles in ecosystems by controlling the evolution and population size of their prey, so changes in their own populations have repercussions throughout the entire food web. Even though not all sharks are apex predators, shark species are often the top predators in coastal marine and estuarine communities, from the tropics to cold-temperate regions, as well as in deep-sea communities to at least 2 miles (3 km) below the ocean surface.

Generalized ecological pyramid of biomass for a marine ecosystem

In this version, the apex predatory shark is pictured on the fifth trophic level. Mesopredatory sharks are one level below, and are potential prey for the apex predator. Biomass of each level increases roughly tenfold as you move toward the base. Overall, sharks are considered tertiary consumers, with a mean trophic level of more than 4. Organisms in the lower two trophic levels are not drawn to scale.

TROPHIC LEVEL

Level 5
Level 4
Level 3
Level 2
Level 1

DISTINGUISHING FEATURES

Sharks, along with their close relatives the batoids (rays and skates) and chimaeras (ghost sharks), belong in the taxonomic class Chondrichthyes, which translates to "cartilaginous fishes," a reference to the principal material comprising their skeleton. Among the nearly 74,000 species of living vertebrates, bone wins outright as the main structural component, making up the skeletons of the bony fishes (around 35,000 species), amphibians (8,400 species), non-avian reptiles (11,700 species), birds (or, more correctly, avian reptiles; 11,200 species), and mammals (6,600 species). Cartilage also forms the skeletal framework of a few "bony" fishes (coelacanths, sturgeon, paddlefish, and bichirs) but is the exclusive component in the 1,245 species of chondrichthyans, a group that breaks down into about 543 species of sharks, 651 batoids, and 53 chimaeras. These numbers increase as scientists discover new species. For example, 40 years ago there were only 342 known species of shark.

In addition to possessing cartilage, sharks have a number of other characteristics that distinguish them from bony fishes. For one, they have 5–6 external gills slits on either side of the head. In batoids, which evolutionarily split from sharks more than 250 million years ago, the 5–7 gill slits are present, but on the underside of the body. Bony fishes have a single bony operculum covering their gills. Another external distinguishing feature is the heterocercal caudal (tail) fin of sharks, in which the upper lobe is longer than the lower lobe. Both lobes of the caudal fin of bony fishes are characteristically equal in length.

WHAT IS A SHARK?

Distinguishing characteristics of sharks

Clockwise from top: Simple one-piece chondrocranium (cartilaginous skull); ampullae of Lorenzini among dermal denticles (scales); jaws showing serial replacement of teeth; between five and seven external gill slits; ceratotrichia (unbranched fin rays); dermal denticles (also called placoid scales); claspers (male only); and cartilaginous skeleton (shown inside the upper lobe of the heterocercal caudal fin)

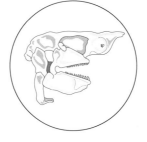

Chondrocranium

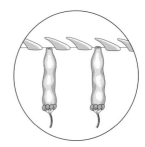

Ampullae of Lorenzini

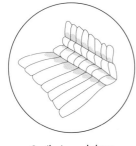

Cartilaginous skeleton

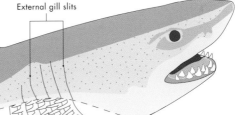

External gill slits

Ceratotrichia

Serial replacement of teeth

Claspers

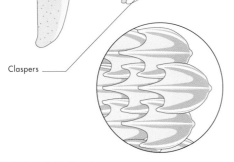

Dermal denticles

← The head of a juvenile Scalloped Hammerhead (*Sphyrna lewini*) in Kane'ohe Bay, Hawai'i (left). Underlying the oddly-shaped head is a laterally expanded cartilage, the material that comprises the skeleton of all sharks. The caudal fin of a Whale Shark (right). This asymmetrical fin, with its elongated upper lobe, which is stabilized by internal vertebral elements, is characteristic of sharks as a group.

THE EVOLUTION AND DIVERSITY OF SHARKS

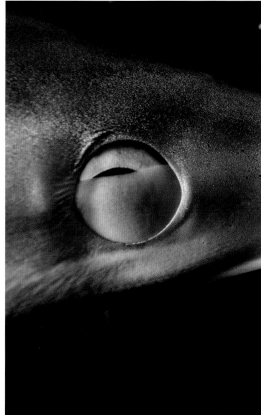

The reproductive lives of sharks are another distinguishing feature, as well as being incredibly fascinating. Sharks differ from the overwhelming majority of bony fishes in that they use a method of transferring sperm to the female that is more often associated with terrestrial vertebrates and marine mammals, namely internal fertilization. Male sharks have two claspers—modifications of the inner margin of the pelvic fins—one of which is inserted into the female. This serves as both an anchoring device and conduit for a brew of seawater and semen, which is flushed into the female's reproductive tract.

Additional shark characteristics include tooth-like placoid scales, a protective eyelid (called the nictitating membrane) in some species, an oil-filled liver (which provides additional buoyancy), a spiral-valve intestine (an unusual adaptation for increasing surface area for food absorption), ampullae of Lorenzini (pores on the head that detect extremely minute electrical fields), and soft fin rays (known as ceratotrichia). This last feature has contributed to the demise of numerous species, since the ceratotrichia, when dried and processed, are the essential ingredient in the Asian delicacy shark fin soup, for which demand is still high despite global campaigns to discourage consumption.

↑ The head of a Sand Tiger (*Carcharias taurus*) displaying multiple rows of teeth and ampullae of Lorenzini, electrosensory pores on its snout.

↖ Mating in Whitespotted Bamboo Sharks (*Chiloscyllium plagiosum*), one of a small number of shark species in which mating has been observed (left). One of the claspers of the male—on the right—is shown inserted into the cloaca of the female. Partially-closed nictitating membrane of a Blue Shark (*Prionace glauca*; page 138 – right). This scale-covered protective structure is found in requiem sharks, and similar structures occur in some birds, lizards, frogs, and even a few mammals (such as seals, polar bears, camels, and aardvarks).

NO TEETHING TROUBLES

If you are a collector of shark teeth, one distinguishing shark characteristic feeds your hobby: serial tooth replacement, or the presence of multiple rows of teeth in various stages of development. Shark teeth are not deeply embedded in the jaw cartilage, but rather move on a looser "conveyor belt" of connective tissue. This constitutes yet another way in which sharks have moved in a different evolutionary direction than all other vertebrate predators. When a terrestrial vertebrate predator such as a Leopard (*Panthera pardus*) loses one of its canine teeth, which are deeply rooted in bone, it enters a dangerous state of competitive disadvantage, since starvation is a major cause of mortality among terrestrial and many marine top predators. Sharks, on the other hand, have planned in advance, evolutionarily speaking, for the certainty of tooth loss. When a tooth becomes dislodged (such as after hitting a bone in a prey animal), it, along with the rest of the entire outer row, is replaced when a new row of sharpened, unworn teeth moves into place. A shark may shed a row of teeth every few weeks or so and go through thousands of teeth over its lifetime.

Why are there more kinds of catfish than sharks?

There are 36 families and 543 species of shark alive today, which may seem an impressive number until you consider that there are more than 435 families and 35,000 species of bony fishes. Even when the other members of the class Chondrichthyes, the batoids and chimaeras, are included, the number of species rises to only around 1,300, or 1.7 percent of known living vertebrates. In comparison, at nearly 4,000 species the taxonomic diversity of catfishes, just one group of bony fishes, is about triple the number of sharks, batoids, and chimaeras combined.

REPRODUCTION STRATEGIES

New species often form following some circumstance or event that separates a population of an existing species. A number of explanations exist for the relatively low speciation rate, and hence low diversity, of sharks and their relatives compared to the bony fishes.

For a start, bony fishes have more offspring than sharks—in other words, they are more fecund. Most bony fishes have external fertilization and invest a small amount of resources into a large number of eggs (several million in the case of pollack, for example).

← The egg cases of a cat shark with early embryos developing within. Egg-laying occurs only in small species of sharks; after a few weeks of development, seams in the egg case dissolve, allowing seawater to enter and the gradually toxifying contents to leave.

→ A rare photo of the birth of a Lemon Shark (*Negaprion brevirostris*) in Bimini, Bahamas. Note the umbilical cord still partially attached to the neonate shark and its mother.

These eggs and the resulting hatchlings have a high ability to disperse, especially if they are pelagic and are entrained in surface currents, which may carry them thousands of miles. The wider their dispersal, the greater the odds of diversifying into new species.

Sharks use the entirely different life history strategy of investing resources into a limited number of offspring—in other words, they exhibit low fecundity. Early development of the shark embryo occurs inside the female for those sharks that lay eggs, such as members of the bamboo shark family (Hemiscylliidae), and complete development occurs internally for species such as White Sharks and Blacktip Sharks (*Carcharhinus limbatus*). The advantage of such a strategy is that each individual has a greater chance of survival than a single bony fish hatchling. However, the disadvantage is that the decreased numbers of offspring have a lower propensity to disperse, and thus have a lower chance of forming new species.

Another explanation for the low speciation rate of sharks is that most (more than 300 species) are small, which limits their ability to disperse. Being small makes migrating across the distances required to achieve the geographic isolation necessary for speciation a very risky prospect. Moreover, about 80 percent of shark species are bottom-dwellers, and are thus less likely to disperse great distances. While most bony fishes are also small, they offset this by producing eggs or hatchlings that can readily disperse.

SIMPLICITY OF FORM

Most bony fishes also have a swim bladder, a gas-filled space (think internal balloon) in the body cavity. This acts as a flotation device and allows them to adjust their buoyancy, and hence their vertical position in the water column (see box). Sharks are heavier than water and must swim continuously or live on the seafloor.

Sharks had about 300 million years during which they could have evolved small forms before the bony fishes achieved this, but they did not. With their fancy swim bladders, the bony fishes came along and literally took over the marine and freshwater aquatic world, aided by the evolution of complex habitats such as reef-building corals, seagrasses, and mangroves.

Finally, bony fishes have more complex heads and jaws than sharks. These provided the raw evolutionary material for them to evolve seemingly unlimited mouth types and head shapes, and thus different feeding styles.

In contrast, sharks have stuck with the same basic plan, with few exceptions. The skull of sharks consists of fewer parts than that of bony fishes. Sharks have a solid skull (chondrocranium), two upper jaw elements (palatoquadrates), and two lower jaw elements (Meckel's cartilages). Such simplicity denies evolution the raw material it needs to diversify. To be sure, the basic shark head plan has been an unqualified success, but along with the other features described above, it deprived them of the ability to diversify to the same extent as bony fishes. There will never be a shark with a mouth like that of a seahorse, wrasse, or butterflyfish.

← A Blacktip Shark at Aliwal Shoals, South Africa. Along the US east coast, the species is frequently observed in dense schools of thousands along its migratory route in spring and fall.

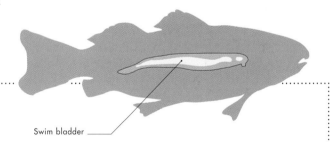

Swim bladder

AN EVOLUTIONARY ADVANTAGE

Sharks do not have a swim bladder. The evolution of swim bladders provided two critical advantages to bony fishes. First, they could save energy by not needing to move continuously. Saving energy is a major driving force for an organism's success, enabling it to devote energy to other survival tasks, as well as mating. And second, they could expand their niches simply by adjusting their buoyancy and using their fins for fine-scale movement. This has allowed them to occupy myriad habitats with more or better food, fewer predators, better protection, or more appropriate environmental conditions. A swim bladder enables bony fishes to exploit the many tight spaces on coral reefs, which account for most of the living space in these ecosystems. In other words, a swim bladder facilitates being small. No sharks on coral reefs are as small as the smallest bony fish species.

THE EVOLUTION AND DIVERSITY OF SHARKS

Are sharks living fossils?

With an evolutionary history dating back about 450 million years, sharks are often called "living fossils," along with crocodiles and ferns. But is this justified?

THE FOSSIL RECORD
Scientists have their hands tied when it comes to deciphering the evolutionary history of sharks. This is because there is a relatively poor fossil record for the shark lineage owing to the inability of most cartilage to fossilize. Overall, shark cartilage, unlike bone, is poorly mineralized. Shark teeth and vertebrae, however, are more mineralized and since these (especially shark teeth) are plentiful in the fossil record, they provide vital insights into shark evolution. Among the sharks alive today, the most primitive are the cow sharks, including the Bluntnose Sixgill Shark (*Hexanchus griseus*; page 38). The teeth of this group distinguish them from all other sharks, and they have been found in sediments dating back 180–200 million years.

Other ways of uncovering the evolutionary history of sharks include imprints of some of the ancestors of modern sharks in fine sediments, as well as anatomical similarities and differences in living sharks. DNA analysis of living sharks is an extremely powerful tool to work out a group's phylogeny, or evolutionary history and relationships.

→ A Broadnose Sevengill Shark (*Notorynchus cepedianus*) in a kelp forest in South Africa. Among sharks swimming today, the most primitive (those with lineage that can be traced back the farthest) are the sixgill and sevengill sharks (order Hexanchiformes).

THE EVOLUTION AND DIVERSITY OF SHARKS

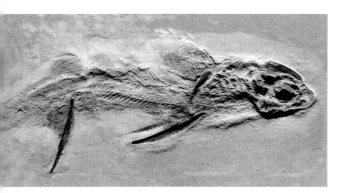

← The fossil of the hybodont shark *Hybodus fraasi*. Hybodont sharks first appeared 360–320 million years ago, and were the dominant sharks for about 100 million years. Some resemble horn sharks, and are perhaps the closest ancestors of modern sharks.

→ A Caribbean Reef Shark (*Carcharhinus perezi*) eating an invasive Red Lionfish (*Pterois volitans*) in a meeting of two species whose ranges should not overlap.

What have these fossils, imprints, and techniques taught us about shark evolution? For a start, the earliest common ancestor of chondrichthyans in general and sharks in particular is still elusive, but paleontologists believe it may have been a group of 440 million-year-old beasts known as acanthodians, or spiny "sharks," which had a cartilaginous skeleton and external gill slits, and looked shark-like. Here is where it gets messy: acanthodians also shared characteristics with modern bony fishes, so it is not entirely clear that they were early shark ancestors.

Sharks likely arose about 450 million years ago, in the Ordovician Period, and persisted through the Devonian (420–360 million years ago). They reached their greatest success in the Carboniferous Period (360–300 million years ago), an era known as the "Golden Age of Sharks." Not only did sharks dominate oceans, rivers, and lakes during that period, but the diversity they exhibited was simply remarkable. Imagine a shark with a large, forward-directed head spine, aptly called the Unicorn Shark (*Falcatus falcatus*). Or one with tooth whorls in the lower jaw, and another with what appears to be scissor-shaped jaws protruding from its mouth. There were 45 known shark families in the Carboniferous, compared with 36 today. Most Carboniferous sharks, along with 90–95 percent of all marine organisms (and 70 percent of terrestrial life), perished in the Permian–Triassic extinction event, also known as the Great Dying. Some chondrichthyan lineages survived, most likely those living in deeper waters. Two such groups were the hybodonts, which later became extinct, and the modern sharks.

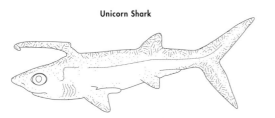

Spiny Shark

Unicorn Shark

Ancient ancestors

The oldest known jawed vertebrates, the acanthodians, or spiny sharks, arose about 440 million years ago and are considered the earliest known shark ancestor. The Unicorn Shark is one of a number of oddball sharks from the Carboniferous Period, known as the "Golden Age of Sharks."

ADAPTED FOR SWIMMING

Many modern sharks are more maneuverable and efficient swimmers than ancestral sharks thanks to an array of adaptations. The pectoral fins became more stabilized, allowing hefty muscles to anchor there. A more elongated snout with a more loosely connected, underslung jaw enabled both heavier jaw muscles and the extremely important modification of greater jaw mobility, and provided room in the braincase for a larger brain. And hardened vertebrae further supported strengthening swimming musculature by providing sturdier points of attachment for heavier musculature. All of these adaptations, plus a few others, allowed sharks to maintain a foothold in their various ecosystems, despite heavy competition from bony fishes and even marine reptiles and mammals. Of these adaptations, the loosened connection between the upper jaw and the braincase was particularly important, since it allowed for the jaw to protrude from the head and created a bigger gape, enabling sharks to eat a wider range of prey items.

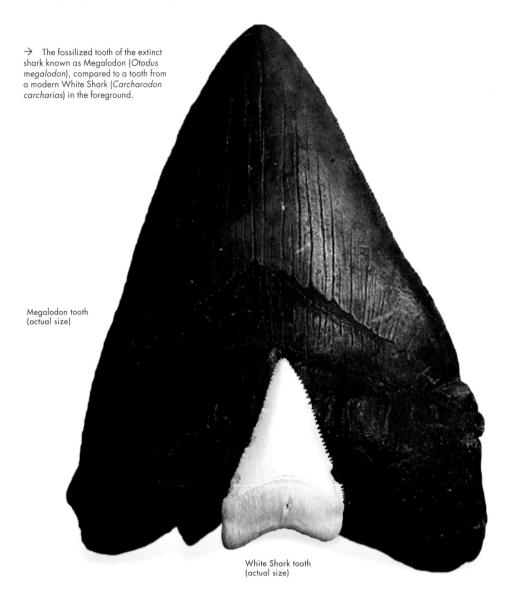

→ The fossilized tooth of the extinct shark known as Megalodon (*Otodus megalodon*), compared to a tooth from a modern White Shark (*Carcharodon carcharias*) in the foreground.

Megalodon tooth (actual size)

White Shark tooth (actual size)

MODERN SHARKS

Also known as neoselachians, modern sharks first arose during the Late Triassic/Early Jurassic Periods, around 200 million years ago, and species similar to many living today arose about 50 million years later, in the Cretaceous Period. Modern sharks diversified into three groups, or superorders: squalomorphs, comprising about 179 extant species of mostly cold-water species, including the cow sharks, frilled sharks, angel sharks, and dogfish sharks; galeomorphs, comprising about 364 species representing a diverse array of sharks, including the horn sharks, carpet sharks, cat sharks, mackerel sharks, requiem sharks, and hammerheads; and batoids, comprising about 651 species of skates and rays.

Sharks are known for their jaws. But the group that sharks split off from were, in fact, jawless, with vacuum-cleaner mouths that sucked up prey from the seafloor. More than 120 species of jawless fishes, known as lampreys and hagfish, have persisted to the present, their diversity limited by the range of prey they could eat without jaws and by competition from the more successful jawed fishes. Jaws originated in evolution from a surprising place: the first set of gill arches, which moved forward and became supported by modifications to the second set. The importance of jaws cannot be overstated, allowing grasping and additional handling of prey.

The ancient shark best known by the public is the beast known as Megalodon (*Otodus megalodon*), the largest predatory fish ever to exist, reaching up to 60 ft (18 m) in length. Megalodon is mistakenly thought of as a direct ancestor of the White Shark. As satisfying as this story would be—that the biggest predatory fish of all time gave rise to the most feared shark alive today—it is apocryphal. Megalodon is an evolutionary dead end and did not give rise to any marine animal living today, not even the White Shark. The latter is thought to have arisen from a relative of the mako sharks and may have helped drive Megalodon's extinction through competition.

Returning to our original question, are sharks living fossils? The answer is a waffling yes and no. Yes, because sharks settled on a successful predatory lifestyle and accompanying body form early in their evolution, and evolution does not usually mess with success. And no, because there have been some impressive modifications, including loosening of the jaw suspension to allow larger gapes and more mobile pectoral fins that increase maneuverability. Whatever the answer, early sharks adopted a predatory lifestyle and evolution has only refined their effectiveness in this role.

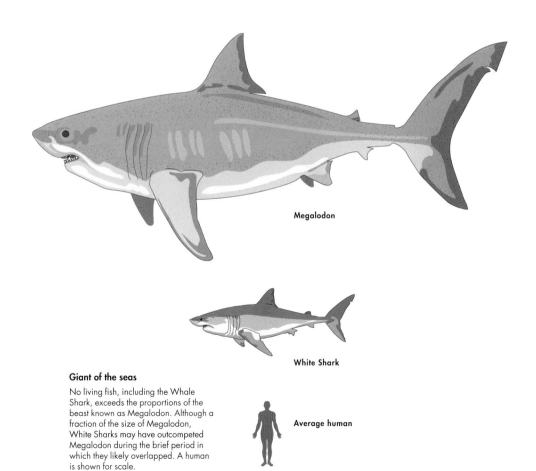

Megalodon

White Shark

Average human

Giant of the seas

No living fish, including the Whale Shark, exceeds the proportions of the beast known as Megalodon. Although a fraction of the size of Megalodon, White Sharks may have outcompeted Megalodon during the brief period in which they likely overlapped. A human is shown for scale.

HEXANCHUS GRISEUS

Bluntnose Sixgill Shark

Giant of the deep

SCIENTIFIC NAME	*Hexanchus griseus*
FAMILY	Hexanchidae
NOTABLE FEATURE	Six external gill slits, cockscomb-shaped teeth in lower jaw, single dorsal fin, big green eyes
LENGTH	15–20 ft (5–6 m)
TROPHIC LEVEL	Top predator and generalist feeder on a variety of mammal carcasses, bony fishes, chondrichthyans, and invertebrates (e.g. cephalopods)

The Bluntnose Sixgill Shark is probably the largest shark you know the least about. With a total length exceeding 20 ft (6 m), the species sits in ninth position on the list of the world's longest sharks, behind the Common Thresher (*Alopias vulpinus*), which seems unfair since the latter owes a large chunk (as much as 50 percent) of its size to its elongated upper caudal lobe. Although the Bluntnose Sixgill is common and widely distributed, in most locations it is found at depths of 1,000–3,300 ft (300–1,000 m), with shallow occurrences in the Pacific Northwest of North America.

The genus *Hexanchus* is the earliest of all living sharks and rays, dating back approximately 190 million years. But the Bluntnose Sixgill Shark is remarkable for more than this and its impressive size. The first notable characteristic is those six gill slits, a feature found in only eight species of sharks, plus a single batoid (most sharks have only five). Two closely related sharks have seven gills. The addition of one or two extra gill arches may reflect the early evolution of these species in oxygen-deficient deep-sea waters.

Other notable characteristics include the large green eyes, specialized for seeing in the dim blue-green light of the species' deep-sea environment. The eyeball also has muscles that pull it back into its socket as a protective measure. The distinctive wide, multi-cusped teeth are an adaptation for sawing pieces of soft tissue out of the carcasses of large animals such as whales, which the sharks may scavenge from the seafloor. Surprisingly, the jaws are poorly calcified and are thus relatively weak, but the weakness has a function in that the jaws can bend across the body of large prey. The sharks also have only a single dorsal fin, which is helpful as they twist or spin their bodies to carve out huge chunks of flesh using their mini-sawblade teeth.

Bluntnose Sixgill Sharks are quite fecund, with documented litters of more than 70 at a time. The most extreme example of multiple paternity (that is, more than one father in a single litter, a phenomenon that is surprisingly not uncommon in sharks) was found in a female that washed ashore dead in Puget Sound on the US West Coast. This single shark had 71 pups in its uteri and genetic analysis revealed that these had been sired by as many as nine males.

→ A Bluntnose Sixgill Shark in Seattle, apparently caught by the camera with its mouth full.

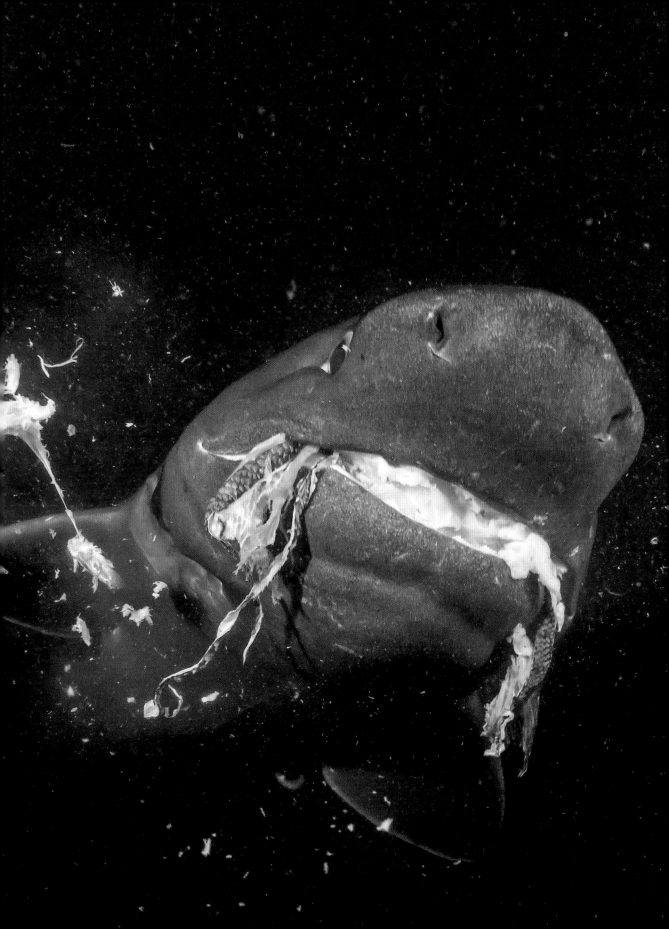

THE EVOLUTION AND DIVERSITY OF SHARKS

GINGLYMOSTOMA CIRRATUM

Nurse Shark

Bottom-dwelling suction feeder

SCIENTIFIC NAME	*Ginglymostoma cirratum*
FAMILY	Ginglymostomatidae
NOTABLE FEATURE	Nasal barbels; first dorsal fin above, or posterior to, pelvic fins
LENGTH	10 ft (3 m)
TROPHIC LEVEL	Nocturnal mesopredator of small fishes and invertebrates

The Nurse Shark is a nocturnal species, so if you've seen one in an aquarium or in the wild, chances are it was motionless, perhaps under a ledge. You may have heard the notion that sharks must continually swim or they will die, in part because they are incapable of pumping water over their gills if not in motion. While this perception is true for many of the more familiar shark species (which would also sink if not moving), benthic sharks and some others are indeed capable of pumping water to irrigate and oxygenate their gills.

The Nurse Shark (not to be confused with the Australian Grey Nurse Shark, *Carcharias taurus*, which is a member of another family) is an inshore benthic (bottom-associated) species that grows to about 10 ft (3 m) in length. It is widely distributed in tropical and subtropical coastal ecosystems. The species is remarkable in that it has extremely powerful jaw musculature, which it uses to suction feed on a diet of mainly small fish, crustaceans, and mollusks. If a hungry Nurse Shark encounters a Queen Conch (*Aliger gigas*), a large snail with a thick shell that deters most predators from the notion of crushing it, the shark positions its mouth over the snail's shell opening and applies suction strong enough to extract and inhale the soft animal from its protective shell—an act that you would be incapable of achieving with your hands. It then proceeds to macerate the snail through a series of bite-and-spit maneuvers that result in a sufficiently softened, bite-sized morsel that is suitable for swallowing.

Highly evolved feeding technique
Underside of the head of a Nurse Shark expanding the space inside its mouth. Doing this creates suction, which it uses when foraging. From left: the mouth is closed prior to feeding; the lower jaw drops and the labial cartilages protrude; maximum suction is created as the gape expands.

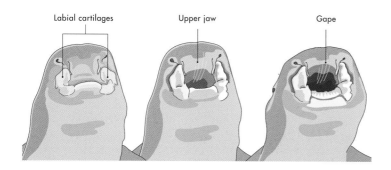

→ A Nurse Shark on a coral reef, Saba, Caribbean. The Nurse Shark is durophagous (eats hard-bodied prey) and uses a suck, crush, spit, and repeat mode of feeding. If provoked, it can attach itself to a person and may be nearly impossible to dislodge.

THE EVOLUTION AND DIVERSITY OF SHARKS

CEPHALOSCYLLIUM VENTRIOSUM

Swell Shark

Shark water balloon

SCIENTIFIC NAME	*Cephaloscyllium ventriosum*
FAMILY	Scyliorhinidae
NOTABLE FEATURE	Flattened head with short snout, spotted body with dark saddles
LENGTH	3.3 ft (1 m)
TROPHIC LEVEL	Nocturnal mesopredator of small fishes and invertebrates

The relatively small Swell Shark (it reaches 3 ft/1 m in length) is not an apex predator, and in fact is named for its predator-avoidance behavior. This sees it making a water balloon out of its body by actively and quickly swallowing seawater to inflate itself. It may also grab its own tail while it doubles in size, making it more difficult for smaller predators to bite. If the shark expands while under a ledge, it is capable of wedging itself there such that it cannot be dislodged. This ability to swallow water and increase in size is unique among the sharks. If a predator does eat a Swell Shark, it may be disappointed by the weakly muscled body of its chosen meal.

Another fascinating adaptation of the Swell Shark, along with its relative the Chain Catshark (*Scyliorhinus retifer*), is that it can glow, or biofluoresce, reflecting bright colors when exposed to certain light. This differs from bioluminescence, which is actual biological light production, as seen in lantern sharks (page 78). However, in both cases the species apparently exhibit a unique pattern of fluorescence or luminescence that could allow them to communicate with other individuals, attract prey, or perhaps camouflage themselves.

Swell Sharks live exclusively in the eastern Pacific Ocean in relatively shallow water (typically at depths of less than 130 ft/40 m). They belong to the family of sharks called cat sharks, which are oviparous (egg-layers). The female Swell Shark produces egg cases with long tendrils, which become wrapped around kelp to hold the egg case in place for the 12 months or so it takes the pup to develop and hatch.

→ A Swell Shark off the California coast. The species belies the reputation of sharks as fearsome, as it is known for its predator-avoidance behavior more so than its predatory prowess.

THE EVOLUTION AND DIVERSITY OF SHARKS

SPHYRNA MOKARRAN

Great Hammerhead
Uniquely adapted head

SCIENTIFIC NAME	*Sphyrna mokarran*
FAMILY	Sphyrnidae
NOTABLE FEATURE	Cephalofoil (head) with straight anterior margin, high first dorsal fin
LENGTH	16 ft (5 m)
TROPHIC LEVEL	Top predator of bony fishes, sharks, and batoids

If any sharks could be considered otherworldly, the hammerheads certainly fit that category. Among the members of the hammerhead family (Sphyrnidae), the most iconic is the Great Hammerhead. Hammerheads are named for their laterally expanded heads, which are called cephalofoils, in recognition of their resemblance to airfoils (airplane wings) and the perception that they provide lift to the shark in the same way wings do for an airplane. The head owes its odd shape to expanded lateral cartilage underneath the skin.

For such an odd and unique shape like the cephalofoil to evolve and be retained in a group of sharks, it must have some overarching adaptive value—in other words, it must help the shark survive in its environment. Determining exactly how this feature confers some advantage to its owner is not an easy task, but it seems the widened head serves as scaffolding that can distribute the head's sense organs over a wider distance. For example, researchers have shown that hammerheads have a wider surface for electroreception and better binocular vision compared to other sharks. The cephalofoil also makes the shark more maneuverable by narrowing its turning radius, but this comes at a cost as the large head also provides lift to the front of the shark. This lift is balanced by having relatively small pectoral fins and an elongated upper lobe to the caudal fin. Finally, hammerheads have been observed pinning down one of their favorite foods, stingrays, to the seabed with their head while they maneuver to bite the prey.

The Great Hammerhead inhabits tropical and warm-temperate seas from the surface to a depth of 900 ft (275 m), and can grow to more than 16 ft (5 m) in length. It produces live offspring that are nourished through an umbilical connection to the mother during development, analogous to placental birth in mammals.

→ A Great Hammerhead highlighted against a colorful Caribbean sky. Their unique head, called a cephalofoil, may confer advantages in the areas of swimming, maneuverability, and locating and handling prey.

THE EVOLUTION AND DIVERSITY OF SHARKS

CETORHINUS MAXIMUS

Basking Shark
Gentle giant

SCIENTIFIC NAME	*Cetorhinus maximus*
FAMILY	Cetorhinidae
NOTABLE FEATURE	Large gill slits, conical snout, caudal keels, small teeth, huge mouth
LENGTH	40 ft (12 m)
TROPHIC LEVEL	Planktivorous

The Basking Shark is second only to the Whale Shark (*Rhincodon typus*; page 144) in the list of the world's largest sharks, but it takes no back seat when it comes to having the largest mouth—its gape can exceed 3 ft (1 m) in width. A shark that can reach 40 ft (12 m) in length and be equipped with a huge mouth could be terrifying, but this giant is actually one of only three species of shark that feed on plankton, and that huge mouth has only a few rows of tiny hooked teeth.

Named after their tendency to "bask" on the ocean surface, Basking Sharks feed actively near the surface, moving through dense patches of zooplankton with their mouth open. Huge volumes of water (and plankton) pass through their gill rakers, which sieve out the plankton and allow the water to exit through extremely large gill slits. Although Basking Sharks are referred to as coastal–pelagic inhabitants of cool-temperate oceans, they have been documented migrating thousands of miles following patches of zooplankton. Some individuals have even been tracked traveling from Massachusetts in the United States to Brazil. Researchers have also discovered that Basking Sharks don't always remain near the ocean's surface, and have tracked them diving to depths of more than 3,000 ft (900 m).

Reproduction in Basking Sharks is similar to that of the Shortfin Mako (*Isurus oxyrinchus*; page 74), with developing embryos relying on unfertilized eggs for nourishment. Basking Shark pups are believed to be 5–6 ft (1.5–1.8 m) long at birth, making them likely the largest of all shark neonates.

Basking Sharks were heavily hunted for their meat, livers (which are rich in oil), fins, and even their skin, for use as a leather called shagreen. As a result, their numbers dropped dramatically and harvest and trade of the species is now heavily restricted in many countries.

→ A Basking Shark in full filter-feeding mode, with plankton-laden water moving into its prodigious gape and exiting its huge gill slits. Staying afloat at the surface while foraging is facilitated by a deceptively fast swimming speed (about 3 ft, or 1 m, per second) and a large, buoyant liver comprising as much as 25 percent of its body weight.

ADAPTATIONS OF SHARKS

Shark anatomy

Like tuna, salmon, and minnows, sharks are fish: aquatic vertebrates with gills for respiration and fins for swimming. As discussed in the previous chapter, sharks and their relatives are distinguished externally from the other major group of fishes, the bony fishes, by having 5–7 external gill slits and an asymmetrical caudal fin. Other external characteristics, however, are similar to those of bony fishes.

SHARK FINS

Sharks have two sets of paired fins, the more forward pectorals and the more posterior pelvics. They also have median fins along the body's midline. These include either one or two dorsal fins, which may have spines on their leading edge, and, in most groups, an anal fin. Some groups—for example, the cow sharks—have lost a dorsal fin through evolution and have only one. Fins serve a variety of functions in sharks, not all of which are apparent and not all of which are the same in every species. What shark biologists do know about the roles of fins comes mainly from a limited number of studies on inanimate models and a few smaller species capable of living in captivity. The latter include the Leopard Shark (*Triakis semifasciata*), Spiny Dogfish (*Squalus acanthias*; page 264), and Whitespotted Bamboo Shark (*Chiloscyllium plagiosum*; page 238), none of which is broadly representative of the wide range of lifestyles of sharks.

One major role fins play is to stabilize the shark when it is in motion, preventing it from rolling (spinning in a circle along the long axis of the body), yawing (moving from side to side), and pitching (moving up and down). The dorsal fins and, to a lesser degree, the anal fin, serve primarily as keels. The pectoral fins control pitch and roll, and possibly yaw, and the pelvic fins also play a role in pitch control.

Fins are also important for maneuverability in sharks. No shark is capable of swimming backward in the way that bony fishes can (dragging a shark backward with a boat, as some fishers are known to do, damages its gills and is often fatal). However, some benthic species such as the Epaulette Shark (*Hemiscyllium ocellatum*) can move backward when in contact with the substrate. While their main propulsion for swimming comes from the caudal fin,

SHARK ANATOMY

OTHER ANATOMICAL ADAPTATIONS

Since all sharks use internal fertilization, males have paired claspers on the inner margins of their pelvic fins (page 26). Inside the shark is the standard suite of vertebrate organs. A four-chambered heart rests inside the pericardial cavity, near the anterior of the shark. In the abdominal cavity are the organs of digestion and reproduction, the liver being the largest of these. In addition to its central role in metabolism, this oil-filled organ adds buoyancy that compensates for the heavy musculature of most sharks.

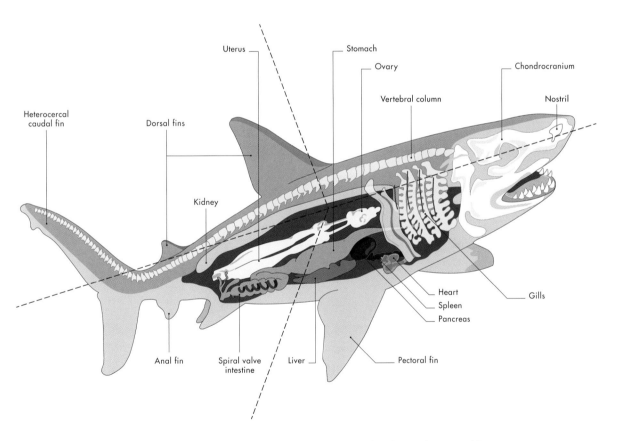

Selected internal anatomical features of a typical female shark

The heart, located slightly behind the gills, pumps blood to the gills, before being distributed to the rest of the body. Food enters the mouth, then traverses the various parts of the digestive tract, including the esophagus, stomach, and spiral valve intestine, before waste is eliminated at the cloaca. The liver, the largest organ in the shark, lies under the digestive organs.

← Two Scalloped Hammerheads (*Sphyrna lewini*) swimming in the Red Sea off Egypt.

ADAPTATIONS OF SHARKS

Epaulette Sharks can crawl along the seafloor and even when emerged between tide pools, using their pelvic and pectoral fins. Skates are also capable of crawling on the seafloor using modifications of their pelvic fins, a phenomenon known as punting. Could these adaptations signify an unrecognized evolutionary trend that may have led to the development of limbs among vertebrates? Unfortunately, no. While limbs indeed developed from fins, it is the lobe-finned fishes, with living relatives that include the lungfishes and coelacanths, that are responsible for the emergence of limbs.

Although the fins of sharks have evolved to become less rigid, they are still stiffer than those of bony fishes. This stiffness, plus the limited flexibility of a shark's vertebral column and the tightness of its skin, limits how tightly some species can turn. That said, species such as juvenile Scalloped Hammerheads (*Sphyrna lewini*), Lemon Sharks (*Negaprion brevirostris*; page 230), and Bluntnose Sixgill Sharks (*Hexanchus griseus*; page 38) are all flexible enough to bite their own tails.

THE SHARK HEAD

The head of sharks houses the sensory organs, including eyes, nostrils, external ear openings (yes, sharks have ears!), and a series of gel-filled pores called ampullae of Lorenzini (page 26). Together with the lateral line, which runs the length of the body on both flanks, these senses enable sharks to interpret their surroundings and respond appropriately. Specifically, they help them detect prey, predators, members of the same species (including potential mates), other organisms, and objects that may be obstacles. They also help sharks orient themselves and even migrate. Benthic and other sharks (such as the dogfishes) have paired, bilateral openings behind the eyes called spiracles (actually modified gill arches) that connect the mouth to the water environment and allow the gills to be irrigated with seawater when the mouth is closed.

↑ A Scalloped Hammerhead amidst a school of fish in waters off Australia.

↖ A Triton Epaulette Shark (*Hemiscyllium henryi*) in the Indo-Pacific Ocean near West Papua, Indonesia. Epaulette sharks live in shallow reef habitats and often crawl on the bottom and even over emergent rocks and coral.

→ A pair of Lemon Sharks (*Negaprion brevirostris*) skim the surface of the water in a synchronous dance off Grand Bahama. Their lateral lines and other senses enable these coordinated movements at close distances.

SHARK ANATOMY

ADAPTATIONS OF SHARKS

The most magnificent fish in the sea

Some of the most fascinating aspects of shark biology are the adaptations that make them such superlative predators. Adaptations refer to aspects of an organism's anatomy, physiology, and behavior that help it survive in its environment and that are hereditary. The Velvet Belly Lanternshark (*Etmopterus spinax*; page 78), for example, uses bioluminescence on its ventral side as a form of counterillumination so that it is hidden from potential predators beneath it. In sharks, paramount among these is their superior ability to locate their prey and catch it, such as the use of sensory barbels and the elongated, blade-like toothed rostrum of the Sixgill Sawshark (*Pliotrema warreni*).

← A Shortfin Mako (*Isurus oxyrinchus*) escapes its watery confines to feed on a piece of bait near Cape Point, South Africa. Its large eyes, ampullae of Lorenzini (pores on the snout), and nostrils all play roles in its superior prey-finding ability.

BUILT TO HUNT

A candidate for paragon of predatory sharks, the ultimate marine predator, is the Shortfin Mako (*Isurus oxyrinchus*; page 74). This species is a blue-water shark, meaning that it lives in the open ocean, with its vast spaces and relative low abundance of prey. The Shortfin Mako has an array of senses, as well as a large brain for collecting and processing sensory information. Once prey—say a Bigeye Tuna (*Thunnus obesus*)—is located, the mako's streamlined body enables it to glide smoothly and stealthily through the water to reach it. To help avoid detection by its prey as it approaches, the shark has countershading, with blue on top and white on the underside. This allows it to blend into both the downwelling surface light when it is viewed from below, and the deep-blue darkness when viewed from above.

THE MOST MAGNIFICENT FISH IN THE SEA

The makings of a high-performance predator

Characteristics that contribute to the predatory prowess of the Shortfin Mako. Clockwise from the top: a relatively large brain; rearward-pointing dagger-shaped teeth; networks of intermingled blood vessels known as retia mirabilia that work to conserve heat and thus warm the body; a heart larger than that of most other comparatively sized sharks; heavy musculature with prominent red muscle fibers (for continuous swimming); and a well-developed suite of senses (lateral line is shown). Also note the large gill slits.

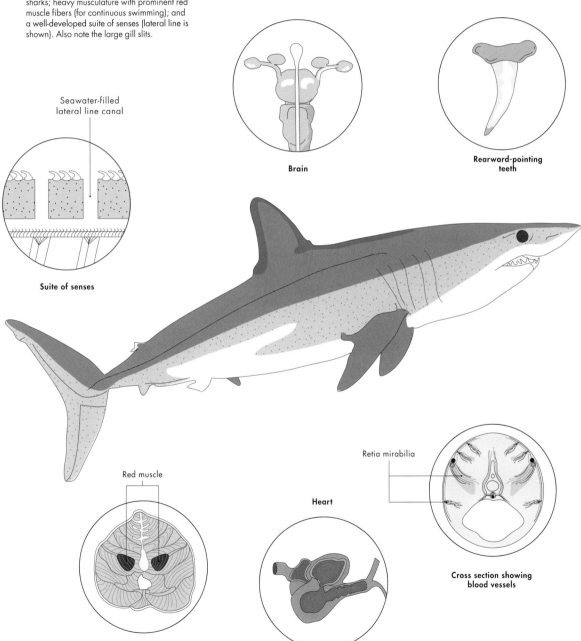

On the Shortfin Mako's approach, its crescent-shaped caudal fin provides the optimal combination of constant thrust and acceleration. And its huge gill slits hide a very large set of gills, which capture essential oxygen from seawater for the body's high-performance aerobic metabolic engine. Once the prey is captured, it will likely be swallowed whole, its escape thwarted by the backward-pointing teeth of the Shortfin Mako's impressive jaw.

METABOLIC ADAPTATIONS

Underlying the Shortfin Mako's adaptations for predation is an additional and extremely unusual feature for a fish: the ability to keep its body warmer than the surrounding water. Most cartilaginous and bony fishes are ectotherms, meaning that their internal body temperature is the same as the environment's. The thermal properties of water make it extremely difficult for aquatic animals to retain their body heat, which is why an air temperature of 70°F (21°C) feels comfortable to us, but water of the same temperature is considered dangerous and will lead to hypothermia in as few as three hours because it draws away your body's heat. Marine mammals successfully keep warmer than the water in which they reside using blubber and fur, neither of which is available to fishes. Fishes are further disadvantaged: their body heat is readily lost to the water when their blood traverses their thin-walled gills to obtain oxygen.

Two groups of fishes—the five species of mackerel sharks (which include both species of mako, White Shark (*Carcharodon carcharias*; page 106), Porbeagle (*Lamna nasus*; page 110), and Salmon Shark (*Lamna ditropis*; page 136), as well as the tunas and members of five other fish families—have evolved specializations to retain some of the heat that would

← A Shortfin Mako in its splendorous entirety, a high performance beast of a shark as magnificent as it is deadly to its prey.

otherwise be lost to their environment. They do so using an elegant engineering principle called countercurrent exchange. In simple terms, the blood that is warmed by heat generated by muscles deep in the body as they work to propel the shark moves in a series of vessels that run straight from the interior near the backbone to the surface. These vessels are juxtaposed with other blood vessels that carry cooler blood from near the outermost parts of the shark straight back to the interior. As the warmed blood moves adjacent to the cooled blood, the heat is transferred from the former to the latter, trapping it inside the shark's body. Through this adaptation, a Shortfin Mako's body temperature can be elevated by as much as 14°F (8°C) over that of its environment.

This metabolic machinery, combined with a high-performance cardiovascular system, enables the Shortfin Mako to have more powerful muscles and thus swim faster. A warm body also allows it to move more independently of seawater temperature than ectothermic sharks, and it improves the acuity of the shark's senses, in part by speeding up the nerve impulses from the sense organ to the central nervous system, where the sensory signal is also processed faster.

Alas, there is no such thing as a free lunch, and maintaining a warm body in seawater comes at a cost. The metabolic furnace that provides the surplus of heat—specifically the muscles—must be fed constantly. But we're talking about a magnificent, super-efficient beast of a fish here, so eating up to 4 percent of its body weight each day is not a chore.

Behind the myth of shark senses

You may have heard sharks being referred to as "swimming noses," a reference to their superior sense of smell. Like many assertions about sharks, there is both truth and myth in this statement. The nose of sharks—paired nostrils on the underside of their snout—is sensitive to smells associated with certain chemicals at relatively low concentrations, such as proteins emanating from a wounded fish. Two caveats about shark smell are in order here: first, in most cases a shark's sense of smell is no more acute than that of an ecologically similar bony fish; and second, smell is only part of the shark's sensory repertoire, and its senses exist in a hierarchy.

THE SENSITIVE TYPE

The ocean is full of sensory information. A shark may use some of this information to find prey, avoid predators, find a mate, avoid unhealthy environments, navigate around obstacles, or even migrate over great distances.

Let's consider locating prey—in this example, an injured fish that has escaped from another predator or that was speared by a diver. Sharks are most sensitive to irregular, low-frequency sounds, such as those emitted by a struggling fish. These sounds might register in the inner ears of a shark as much as 2 miles (3 km) distant.

As the shark orients to and swims toward these sounds, it might pick up the scent trail from the bits of tissue and fluid emanating from the wounded fish about 1,000 yd (1 km) away, providing it is downstream of the prey. As the shark swims in and out of the odor corridor, it detects the strongest scent and moves in that direction. Then, when it has closed the distance between itself and the targeted prey to about 65 ft (20 m), and assuming the water is clear, it may use its vision to locate the prey. This is fortunate, since smell and sound can be confusing sensory signals close to the prey.

← A Sand Tiger (*Carcharias taurus*) noses its way through a massive baitball of tiny fish off the coast of North Carolina. These baitballs sometimes conceal the sharks in their midst and provide a measure of stealth to the predator.

Hierarchy of a shark's sensory system

An overview of the senses used by a shark to locate prey based on distance from the stimulus (sound, smell, water movement, sight, bioelectrical field, and taste).

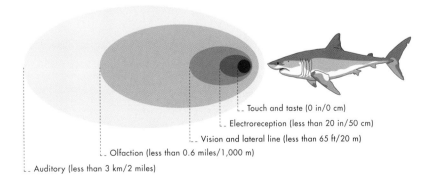

Touch and taste (0 in/0 cm)
Electroreception (less than 20 in/50 cm)
Vision and lateral line (less than 65 ft/20 m)
Olfaction (less than 0.6 miles/1,000 m)
Auditory (less than 3 km/2 miles)

ADAPTATIONS OF SHARKS

← A White Shark (*Carcharodon carcharias*) feasts on a marine mammal carcass, a fortuitous find that provides high-quality nutrition to feed the shark's demanding metabolic machinery. The shark's eyes have reflexively rolled back, exposing the white, tough, fibrous protective sclera.

→ Two juvenile Port Jackson Sharks (*Heterodontus portusjacksoni*), a horn shark common in southern Australia. Should a predator such as an angel shark (*Squatina*) conclude that one would make a delicious snack, it should think again: the stout spines on the dorsal fins make all nine species of horn sharks unpalatable, even dangerous, to most predators.

The aromas of a great restaurant and the clinking and sizzling sounds emanating from its kitchen may draw you inside, but once there, these signals alone won't guide you to the source of the food.

Vision will take a shark to within 1.5 ft (50 cm) of the prey, and its lateral line will also detect the struggling fish. But the sensory environment can still be confusing at close distances, especially when the shark opens its mouth immediately in front of the prey. At that point, movement of its head may enlarge its natural blind spot or, in many sharks, protective eyelids called nictitating membranes rise to cover the eyes. In White Sharks, the entire eye rolls 180 degrees in the socket, exposing the whites. Again, sharks have this situation covered, since their ampullae of Lorenzini, a series of gel-filled pores on shark heads (page 26), detect the minute electrical field of the hapless prey, and gulp, it is history.

What happens if the shark makes a mistake and the prey's tissues are poisonous? Sharks have this situation well in hand too, since taste buds lining their mouth sample the fish and can instantaneously decide if the prey is suitable. If not, it is rejected. These taste buds may account for bite-and-release shark interactions where, say, a Blacktip Shark (*Carcharhinus limbatus*) mistakes a swimmer's feet for a swimming fish, only to be informed by its own taste buds to release and move on.

HEARTLESS HUNTERS?

It is true that the heart of sharks, as well as bony fishes, is quite small compared to that of mammals. Our heart weighs about 0.7 percent of our total body weight, whereas the heart of a shark weighs from about 0.1 percent of its body weight (Horn Shark, *Heterodontus francisci*) to more than double that for a Shortfin Mako or White Shark. A human's heart weighs about 2.5 times that of the White Shark.

The heart's function is to pump blood to a series of vessels that distribute the blood to the parts of the body where it is needed most at any particular time. The relative size of the heart correlates most directly with an animal's activity level and metabolic rate—

in other words, how quickly it uses energy and how much energy it uses (typically measured by the rate of oxygen consumption). The more sedentary Port Jackson Shark (*Heterodontus portusjacksoni*; page 76), which typically rests during daylight hours on the seafloor and moves episodically and slowly at night, requires a smaller heart than sharks at the other end of the metabolic spectrum, such as the Shortfin Mako.

Two other features distinguish the shark heart from that of other vertebrates, including bony fishes. First, the shark heart lies within a capacious pericardium that has relatively inflexible walls (our pericardium, in contrast, envelops the heart more tightly and is flexible). When the largest and most muscular of the shark's four cardiac chambers contracts, the pericardial walls resist moving in, and suction is developed within the pericardial space (according to basic physics, if the volume in a fixed space declines, so must the pressure). This suction pulls blood back into the heart from the posterior. Cardiac suction occurs in all vertebrates, but not to the extent that it does in sharks.

The second distinguishing feature of the shark heart is the presence of a pressure-relief valve in the wall of the pericardium. Only a few oddball fish have this adaptation, which allows the heart to expand when it needs to pump more blood by squirting some of the pericardial fluid into the abdominal cavity. And this adaptation helps prevent some sharks from dying when the heart is injured, an example being when a Port Jackson Shark eats a sea urchin and one or more of its spines pierce the shark's heart and cause it to bleed. Without a way to vent the blood in such a situation, pressure in the pericardium would increase and likely kill the shark. In another example, one of the authors caught a perfectly healthy Tiger Shark (*Galeocerdo cuvier*; page 232) with a 4 in-long (10 cm) catfish spine sticking out from its pericardium. The shark had eaten the catfish and the spine had passed from its esophagus into the pericardium, presumably through part of the heart, and then eventually worked its way out.

ADAPTATIONS OF SHARKS

↑ A Blacktip Reef Shark (*Carcharhinus melanopterus*) in Maldives. This species is a common mesopredator on shallow Indo-Pacific reefs; its fins appear to have been dipped in India ink. It is often confused with its larger cousin, the Blacktip Shark (*C. limbatus*).

↗ Mesmerizing photo of sharks feeding on a bait ball in the breaking ocean waves off Carnarvon, Western Australia.

BEHIND THE MYTH OF SHARK SENSES

ADAPTATIONS OF SHARKS

Unparalleled reproductive adaptations

The suite of shark adaptations related to reproduction will likely far exceed your expectations. They range from the novel mechanism males use to internally fertilize the females' eggs, to the incredible array of modes females employ to nourish their embryos.

SEXUAL DIMORPHISM

Thoughts of sexual dimorphism, or differences in appearance between males and females, likely conjure up images of birds with starkly different plumages (for example, the showy tail feathers of the peacock versus the drab coloration of the peahen) or the presence of large antlers in bull Elk (*Cervus canadensis*) and Moose (*Alces alces*) compared to the antlerless cows. Sharks also show marked dimorphism between the sexes, the most obvious being size dimorphism, with females often being much large than males. These differences are most extreme in species that give birth to live young, since the female must carry the offspring, and are less so in species that lay eggs.

↑ A pregnant Bull Shark (*Carcharhinus leucas*) off Playa Del Carmen, Mexico. Up to 13 offspring are born after 12-months' gestation.

← Nurse Sharks' (*Ginglymostoma cirratum*) courtship in the shallows. A male has grasped the pectoral fin of the female, which may be followed by mating. Shark biologists Wes Pratt and Jeff Carrier were the first to describe social structure and mating behavior in the species.

The most obvious sexually dimorphic feature in sharks are the claspers mentioned earlier. These tubular extensions of the pelvic fins, stiffened by calcified cartilage, are present only in males and provide the mechanism for transferring sperm into the female for egg fertilization. Males have two claspers, but only one is inserted into the female's cloaca during mating (which one depends on the position of the male relative to the female). Nerves stimulate the clasper to rotate forward, and once it is inserted, the end splays open, anchoring it into the female during mating. Depending on the species, the splayed clasper may display a variety of hooks, spikes, or lobes to assist in anchoring.

Sharks have other sexually dimorphic characteristics related to the mating process itself. Males often bite the fins or flanks of the female during copulation, causing visible mating wounds. To combat this, female sharks typically have thicker skin than males. For example, the skin on the flank of a female Blue Shark (*Prionace glauca*; page 138) is three times as thick as the male's. Smooth-hounds and other sharks that feed mostly on invertebrates often have flattened teeth for crushing prey. But males in these species have pointed cusps on their teeth, to assist in holding onto the female during mating.

ADAPTATIONS OF SHARKS

MODES OF EMBRYONIC DEVELOPMENT

Once mating has taken place and the eggs have been fertilized, female sharks employ more modes of nourishing their embryos than any other animal group. Whereas birds and amphibians use only one mode of embryonic development (egg-laying) and mammals employ a mere two (placental live birth and a few egg-layers), there are as many as 10 modes in sharks. In all vertebrates, the yolky part of the egg forms a yolk sac that is attached to the embryo's digestive tract and provides its primary source of nourishment, at least during early development. However, in sharks the many modes of embryonic nourishment exist on a spectrum, from species that get nourishment only from the yolk sac (lecithotrophy, or yolk-feeding) to those that get a significant amount of energy directly from the mother (matrotrophy, or mother-feeding). Cat sharks and horn sharks are egg-layers. In these species the fertilized egg is enveloped in a horny casing that is expelled from the mother and most development occurs outside of the mother using only the attached yolk sac for nourishment. Once the yolk is depleted, the baby shark breaks free of its egg capsule.

All other shark species have some form of live birth. Sharks have paired uteri and, in most, embryo development in both. Sawsharks, angel sharks, cow

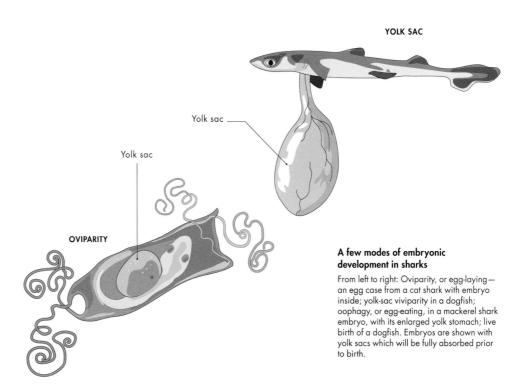

A few modes of embryonic development in sharks

From left to right: Oviparity, or egg-laying—an egg case from a cat shark with embryo inside; yolk-sac viviparity in a dogfish; oophagy, or egg-eating, in a mackerel shark embryo, with its enlarged yolk stomach; live birth of a dogfish. Embryos are shown with yolk sacs which will be fully absorbed prior to birth.

UNPARALLELED REPRODUCTIVE ADAPTATIONS

sharks, dogfish sharks, and other relatives retain the fertilized eggs internally, often with multiple eggs wrapped in a single envelope. In these species, most or all of the energy needed by the developing embryos comes from the yolk sac, with little to no nourishment provided by the mother. In mackerel sharks, including the White Shark, the mother ovulates huge numbers of unfertilized eggs to feed her developing offspring (called oophagy, or egg-eating). After the yolk sac is depleted, the embryos gorge themselves on these eggs until their stomachs become extremely distended. This represents a significant energetic investment from the mother. One species, the Sand Tiger (*Carcharias taurus*; page 236), takes this a step further, with the embryos not only eating the ovulated eggs, but the largest one eating all of its siblings in utero (talk about sibling rivalry!).

In the requiem sharks, hammerheads, and many of the houndsharks, the maternal investment goes a step further. Following depletion of the yolk sac reserves, the embryos' yolk stalks attach to the wall of the mother's uterus, forming a pseudo-placental connection analogous to the mode of nourishment used by placental mammals. Through this umbilicus, the female provides the bulk of the developing embryos' nourishment by shunting uterine milk (called histotroph) directly into their digestive tract.

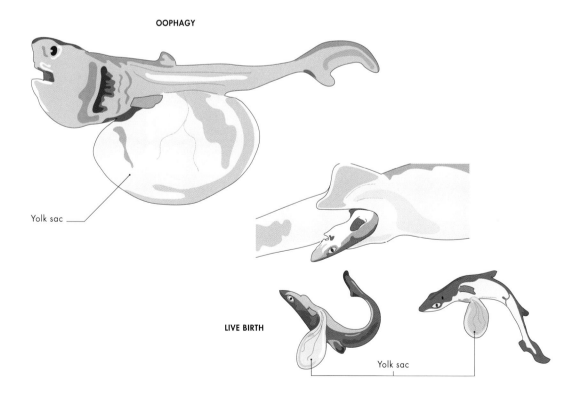

OOPHAGY

Yolk sac

LIVE BIRTH

Yolk sac

FECUNDITY FIGURES

The number of offspring produced by female sharks can vary from as few as one to as many as 300, and even varies among reproductive modes. Egg-laying Small-spotted Catsharks (*Scyliorhinus canicula*) may produce 60 or more eggs in a season, while the Port Jackson Shark produces about 10–16 corkscrew-shaped eggs each year. The known record for shark fecundity is more than 300 in the Whale Shark (*Rhincodon typus*; page 144), a species that essentially holds encased eggs internally. Among the live-bearing yolk-sac species, the Bluntnose Sixgill (*Hexanchus griseus*; page 38) and Prickly Shark (*Echinorhinus cookei*) may produce more than 100 offspring, yet at least one species of gulper shark has only one pup after a gestation period of at least two years. In egg-eating and embryo-eating sharks, fecundity is quite low owing to the extreme level of maternal investment, and can range from just two per cycle in Sand Tigers and thresher sharks to perhaps 14 in White Sharks (*Carcharodon carcharias*; page 106). Fecundity in placental shark species also varies considerably, with Whitetip Reef Sharks (*Triaenodon obesus*; page 114) usually producing only 2–3 offspring, although Blue Sharks and Great Hammerheads (*Sphyrna mokarran*; page 44) may produce 40 or more pups.

↑ A cluster of cat shark egg cases. Note the tendrils—curly, wiry threads—that entangle with each other and the bottom, making removal of the egg cases very difficult.

→ An egg from a Port Jackson Shark on a beach in New South Wales, Australia. The eggs are often lodged between rocks, where they are protected and do not typically get washed onto the beach. Females have been observed carrying their eggs in their mouths and positioning them among the rocks.

UNPARALLELED REPRODUCTIVE ADAPTATIONS

ADAPTATIONS OF SHARKS

Shark personality and individual recognition

The study of shark behavior has widened considerably in recent years. Early studies of sharks understandably focused on behaviors associated with shark bites and attacks, and this remains a popular subject of inquiry.

PREDATORY AND DEFENSIVE BEHAVIOR

One of the earliest notable observations was made in the 1970s by shark researcher Don Nelson. Divers in submersibles moving aggressively toward Grey Reef Sharks (*Carcharhinus amblyrhynchos*) in the wild repeatedly and predictably induced the sharks to move into a stylized series of postures. The more aggressive the divers and the fewer the escape options available to the sharks, the more pronounced the sharks' defensive and warning display, which often resulted in strikes on the submersible. Nelson's research was the first to correlate shark attacks on humans with a threat perceived by the shark, and not predatory behavior. Bull Sharks (*Carcharhinus leucas*; page 196) exhibit similar behaviors, and have been known to attack boats that move too close.

Understandably, studies of White Shark feeding behavior have captured the interest of the public, the media, and researchers. One particularly memorable behavior involves White Sharks ambushing juvenile Cape Fur Seals (*Arctocephalus pusillus*) in South Africa from beneath with an aerial attack in which the shark grasps the unsuspecting prey while launching its entire body out of the water. Overall success rates varied with time of day, but averaged slightly less than 50 percent.

In addition to trying to understand predatory behavior in sharks, recent studies have focused on their social lives, cognitive abilities, navigation, and predator avoidance. For example, researchers have found that juvenile Lemon Sharks (*Negaprion brevirostris*; page 230) in shallow mangrove-lined lagoons form groups at high tide, when predators—usually larger Lemon Sharks—are present. The smaller Lemon Sharks selected similar-sized individuals to hang out with, and even showed a preference for those they were familiar with from previous encounters. Individual recognition has also been shown in other shark species. Sharks of the same species may look alike to you, but not to one another!

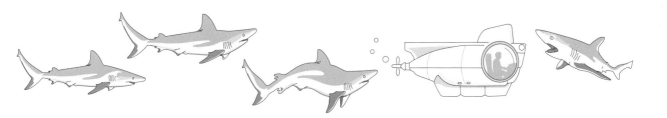

↑ Lemon shark juveniles in their mangrove nursery, Bimini, The Bahamas. Studies there have discovered that juveniles like those pictured exhibit different personalities and can recognize familiar individuals.

← Don Nelson next to his Shark Observation Submersible at Enewetak Atoll, Marshall Islands. The vehicle was designed and built by Nelson and his students and was used for pioneering studies of threat displays in Grey Reef Sharks (*Carcharhinus amblyrhynchos*) in 1978.

Agonistic, or threat display, by a Grey Reef Shark

The unthreatened shark swims normally until it perceives an intruder, in this case a submersible, after which it slows and swims more jerkily while displaying a stylized series of poses warning the intruder that it is not fooling around, which the submersible and diver will soon discover!

LEARNING BEHAVIOR

Sharks are also capable of learning, a fact experimentally verified in the 1950s using adult Lemon Sharks that were trained to bump a target, which resulted in a food reward. Grey Bamboo Sharks (*Chiloscyllium griseum*) are capable of discriminating between shapes, contrasts, and two-dimensional objects. Perhaps most astounding, however, is cognition in the Port Jackson Shark (*Heterodontus portusjacksoni*; page 76), if only because this primitive-looking species spends its days lazily resting on the bottom and does not look like it would be capable of learning. In one set of experiments, Port Jackson Sharks learned to bite at bubbles emanating from an underwater aerator in order to receive a food reward.

It took as many as 30 trials to learn the behavior, but it was retained in some individuals for about 40 days. In the last few years, sharks of several species—including Caribbean Reef Sharks (*Carcharhinus perezi*; page 234) and Sandbar Sharks (also known as Brown Sharks, *C. plumbeus*; page 198)—have learned to arrive early at provisioning stations set up by dive and education operations and at fishing piers in anticipation of being fed.

→ A provisioning (feeding) station for Caribbean Reef Sharks (*Carcharhinus perezi*) in The Bahamas. Individuals have learned to associate the presence of boats and divers with food and arrive early at these sites.

DO SHARKS HAVE PERSONALITIES?

Most surprisingly, sharks have also been shown to exhibit different personalities. Personality refers to behavioral traits that may differ among individuals within a population—for example, risk-taking versus cautious individuals. Shark researchers at the Bimini Biological Field Station in The Bahamas have anecdotally observed differences in boldness (or inquisitiveness) versus timidity in individual Tiger Sharks (*Galeocerdo cuvier*; page 232), but they have also conducted rigorous scientific testing on groups of juvenile Lemon Sharks. They found that some individuals are more explorative in novel surroundings than others; in other words, these sharks show the personality trait of boldness. The specific value of this variation in particular and of personality differences in general is subject to debate. Having a mixture of personalities in a group of sharks may be protective in the long run, perhaps protecting shier sharks from dangers of risk-taking (for example, exposure to more predators) and rewarding bolder sharks with food, potential mates, and so on.

ADAPTATIONS OF SHARKS

ISURUS OXYRINCHUS

Shortfin Mako
Superlative predator

SCIENTIFIC NAME	*Isurus oxyrinchus*
FAMILY	Lamnidae
NOTABLE FEATURE	Conical snout with protruding, backward-pointing dagger teeth; enormous gill slits; crescent-shaped tail; brilliant blue coloration on top
LENGTH	13 ft (4 m)
TROPHIC LEVEL	Top predator, opportunistically consuming a variety of marine mammals and fishes, including Swordfish (*Xiphias gladius*), tuna, and marlins

Thanks to its spectacular coloration (brilliant blue or purple on top, white on the bottom) and equally impressive adaptations as a high-performance predator, the Shortfin Mako (its official common name) could make claim to being the most magnificent beast in the sea. In *The Old Man and the Sea* (1952), Ernest Hemingway wrote that this shark was "built to swim as fast as the fastest fish in the sea and everything about him was beautiful except his jaws ... This was a fish built to feed on all the fishes in the sea, that were so fast and strong and well-armed that they had no other enemy."

The Shortfin Mako is found globally in temperate and tropical latitudes, and grows to more than 13 ft (4 m) in length. It is easily identified by its conical snout and jaw full of backward-pointing dagger-shaped teeth that protrude even when the mouth is closed. In addition, it has enormous gill slits, relatively short pectoral fins, and a crescent-shaped tail with an upper lobe that is about the same size as the lower lobe.

Ancestors of the Shortfin Mako and its close relatives—including the White Shark, Longfin Mako (*I. paucus*), Porbeagle, and Salmon Shark—arose about 45–65 million years ago. These ancestral species then embarked on an evolutionary journey that ultimately separated them from all of other sharks and made the Shortfin Mako and its relatives the superlative predators they are today. The adaptations accounting for their predatory prowess include an almost perfectly streamlined body, keels on the caudal fin, and especially an elevated body temperature. The last of these adaptations is highly unusual among fishes, since the heat-trapping properties of water make it very difficult for an aquatic organism to be warmer than its surrounding environment. Higher body temperatures increase the force of contraction of muscles and enable more powerful swimming, and allow endothermic sharks to maintain their sensory acuity in cold waters. A Shortfin Mako can cruise over 20 mph (32 kph) and is capable of bursts at or above 45 mph (75 kph).

Makos give birth to live young. As embryos, Shortfin Makos are oophagous, that is, during development they eat the ova that the mother continues to produce during pregnancy. The embryos store the ova in their protruding guts, which are known as egg, or yolk, stomachs.

→ An inquisitive Shortfin Mako off Hawai'i. Shortfin Makos represent the apex of design as ultimate predators, with a suite of adaptations enabling high-performance predation.

ADAPTATIONS OF SHARKS

HETERODONTUS PORTUSJACKSONI

Port Jackson Shark

Nocturnal mesopredator that crushes its prey

SCIENTIFIC NAME	Heterodontus portusjacksoni
FAMILY	Heterodontidae
NOTABLE FEATURE	Short head, piglike snout, high ridges over the eyes, spined dorsal fins
LENGTH	5.4 ft (1.65 m)
TROPHIC LEVEL	Mesopredator of sea urchins, mollusks, crustaceans, fishes

How can you not love a shark with a piglike face? Named after Port Jackson (Sydney Harbour), this species is found along the coast of southern Australia. It is found in rocky coastal habitats, where it is nocturnally active and hides in crevices or among rocks by day.

Let's start with some *nots*. The Port Jackson Shark is not the fastest shark in the sea. Its body shape, with a large head preceding a stout body and large fins, discourages hydrodynamic efficiency; that is, it disrupts the flow of water around it, leading to increased turbulence and greater frictional drag, the enemies of easy swimming. Thus, the Port Jackson Shark moves sluggishly, if at all, when active at night.

Unlike many sharks, the Port Jackson Shark does not have a mouth full of impressively large teeth. Since it feeds mostly on benthic invertebrates, including hard-shelled species, it is best served by small cusped teeth at the anterior to grasp its prey, and molariform teeth surrounded by hypertrophied jaw muscles in the rear to crush it (hence the genus name *Heterodontus*, meaning "different teeth"). Anterior bite force for the species has not been measured, but in the closely related Horn Shark it was only 37 lb (163 N), although maximum posterior bite force on the rearmost molariform teeth was 89 lb (382 N). These figures are low compared to measurements for iconic species such as White Shark and Bull Shark, but relative to the small mass of the Horn Shark, they are some of the highest among sharks.

Port Jackson Sharks have been observed segregating sexually in their natural habitat. They lay individual corkscrew-shaped eggs that are adapted to become lodged between rocks for protection and also so that they do not get washed onto beaches. Port Jackson Shark mothers have been observed carrying the eggs in their mouths and positioning them among rocks.

→ The Port Jackson Shark, a type of bullhead shark, with a face only a mother could love, in New South Wales, Australia. Its truncated snout facilitates eating hard-bodied invertebrates, which they suck into their mouth.

ADAPTATIONS OF SHARKS

ETMOPTERUS SPINAX

Velvet Belly Lanternshark

Bioluminescent deep-sea mesopredator

SCIENTIFIC NAME	*Etmopterus spinax*
FAMILY	Etmopteridae
NOTABLE FEATURE	Two dorsal fins with spines, no anal fin, bioluminescent photophores on underside
LENGTH	18 in (45 cm)
TROPHIC LEVEL	Mesopredator on small bony fish, squid, shrimp and, as juveniles, krill

The poetically named Velvet Belly Lanternshark has a black underside that resembles velvet and is distinct from its brownish flanks. It is one of the most abundant deep-water sharks in the northeast Atlantic, reported at depths to 8,200 ft (2,500 m), although more typically found in the 660–1,600 ft (200–500 m) range. The lantern shark family, which currently comprises 51 species, is characterized by the presence of light-emitting organs called photophores, which make the sharks bioluminescent. These are mostly concentrated on the ventral side of the shark, but in some species they occur in a band on the flanks and caudal peduncle.

Bioluminescence is very common among deep-sea organisms and in the Velvet Belly Lanternshark is thought to function in counterillumination—that is, it produces blue-green light that mimics the frequency of the ambient downwelling surface light. This conceals the Velvet Belly Lanternshark from predators swimming below. Like other deep-sea sharks, the eyes of this species are green, which allows it to maximize vision in the dim blue-green light in its habitat.

In spite of its common occurrence, the species is not well known, much like most deep-sea sharks. Its mode of embryonic nutrition is yolk-sac viviparity; that is, young are born live but there is no additional nutrition from the mother while developing in the uteri. Litters of 6–20 have been reported, with a gestation period of one or two years and a reproductive cycle of 2–3 years. The species is among the most abundant deep-sea sharks and the largest deep-sea shark bycatch in commercial fisheries, and is considered Near Threatened by the International Union for Conservation of Nature (IUCN).

→ A Velvet Belly Lanternshark puts on a bioluminescent light show in deep water off Norway.

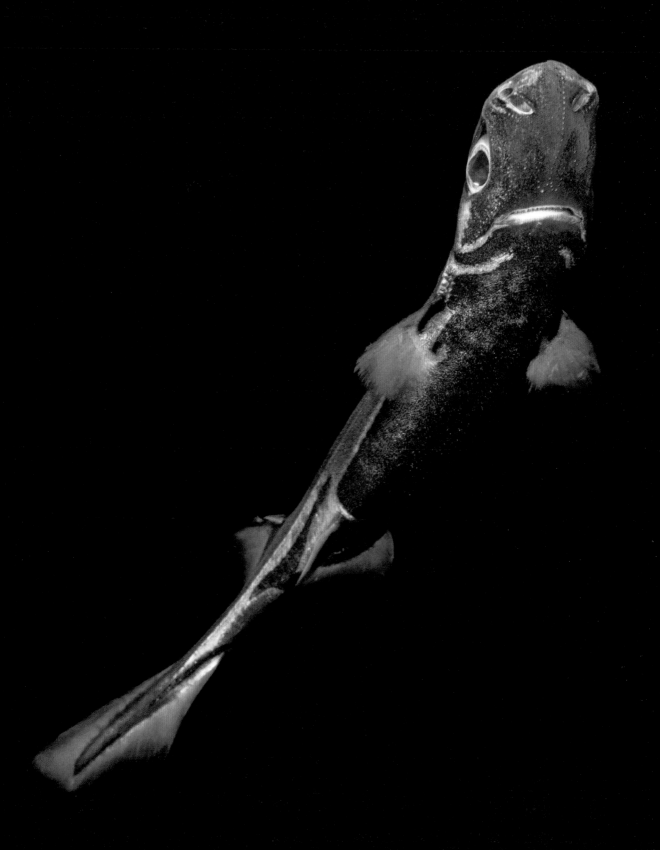

ADAPTATIONS OF SHARKS

HEMISCYLLIUM HALMAHERA

Halmahera Epaulette Shark

Carpet shark occasionally found in tide pools

SCIENTIFIC NAME	*Hemiscyllium halmahera*
FAMILY	Hemiscylliidae
NOTABLE FEATURE	Long tail; two spineless equally sized dorsal fins set far back on body; brown with clusters of brown or white spots
LENGTH	3.3 ft (1 m)
TROPHIC LEVEL	Nocturnal mesopredator on small bony fishes and benthic invertebrates

One of 15 species of long-tailed carpet sharks, the Halmahera Epaulette Shark is slender with a very long tail. Epaulette sharks are named for the black spot outlined in white behind the pectoral fins, which resembles a military epaulet and may function as an eyespot to make predators think twice about eating it. It is found in Indonesia, where it inhabits coral reefs and tide pools. This species is capable of crawling over emergent rocks and coral.

Some isolated tidepool habitats (also called coral reef flats) inhabited by this and other epaulette sharks are subject to nocturnal hypoxia or even anoxia. This is when the organisms, plants, and animals living in the pool use up most or all of the dissolved oxygen (hypoxia and anoxia, respectively) during evening low tides, before it is restored by algal photosynthesis in daylight or by flushing with seawater during subsequent high tides. Since dissolved oxygen is an absolute requirement for sharks and other aerobic life, reduced levels could be deadly. However, epaulette sharks have evolved increased physiological tolerance to even extreme hypoxia, diverting blood flow to the brain, an organ extremely sensitive to the deprivation of oxygen.

These same shallow tide pools may also reach temperatures of 93°F (34°C) during the day, 5–7°F (3–4°C) higher than nearby deeper reef areas. These high temperatures lead to reduced growth rates and even death in juvenile epaulette sharks, which avoid this fate through behavioral thermoregulation—in other words, they select to move to cooler waters. Adult epaulette sharks have a high critical thermal maximum (the highest temperature they can tolerate and still survive), but they must also move if the heat becomes too much.

→ The Halmahera Epaulette Shark, also called the Halmahera Walking Shark, from Moluccas, Indonesia. Both the musculature and skeletal structure of the pectoral and pelvic fins are modified for crawling, in addition to swimming.

ADAPTATIONS OF SHARKS

PRISTIOPHORUS CIRRATUS

Longnose Sawshark

Evolutionary oddity

SCIENTIFIC NAME	Pristiophorus cirratus
FAMILY	Pristiophoridae
NOTABLE FEATURE	Rostrum flattened into toothed saw, teeth alternate between large and small and are conical, prominent elongate barbels on rostrum, anterior to nostrils, flattened body
LENGTH	4.6 ft (1.4 m)
TROPHIC LEVEL	Mesopredator of small fishes and invertebrates

Not to be confused with the sawfish, which are batoids, this species (and the nine others in its family) is a shark, albeit a very unusual and small shark. Sawsharks are easily distinguished from sawfish in that they have both lateral barbels and alternating teeth of different sizes on their saw.

All members of the sawshark family have a toothed rostrum that is elongated into a saw. The rostrum both houses sensory organs and acts as an offensive and defensive weapon, and is likely used in mating. The sharks cruise along the seabed and use the chemosensory barbels and electrosensory ampullae of Lorenzini on the saw to detect prey buried in the sediment. They then hit victims with side-to-side swipes of the saw, crippling them before grasping them with their cusped, non-flattened teeth. Like sawfish, sawsharks cannot replace lost rostral teeth. A sawed rostrum has evolved only three times: in the sawsharks, the sawfish, and the ganopristids, a group of batoids that went extinct about 150 million years ago. There are three species of sawshark with six gills, the only species other than cow sharks with this characteristic.

The Longnose, or Common, Sawshark and its relatives are all viviparous; that is, they give birth to live young. You might wonder whether carrying young equipped with small but still potentially dangerous saws could harm the mother. The answer is no, since the teeth are conical and not bladelike as in the batoid sawfish, and they also lie flat against the blade until birth. In sawfish, a protective collagenous sheath covers the sharp rostral teeth in the embryo, dissolving shortly after birth.

The Longnose Sawshark is listed as Least Concern by the IUCN. It has a limited distribution, in the eastern Indian Ocean off Southern Australia and nearby areas to depths of 1,970 ft (600 m).

→ A Longnose Sawshark in Jervis Bay, New South Wales, Australia. The barbels extending from the rostrum have been found to be largely tactile, and do not have taste buds or other sensory structures.

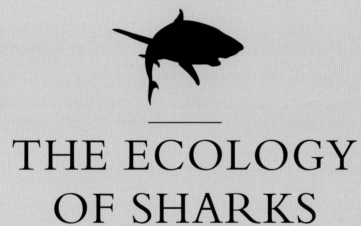

THE ECOLOGY OF SHARKS

THE ECOLOGY OF SHARKS

Ecological roles of sharks

Sharks are an ecologically diverse group, occupying such different habitats as the open ocean, deep sea, estuaries, rivers, and temperate and tropical coastlines. What roles do sharks play and how important are they in their ecosystems? What are the consequences of their loss? What, how often, and how much do sharks eat? What predators eat sharks? Do sharks remain in one habitat year-round, or do they migrate? How long do sharks live? How fast do sharks grow? This chapter gives answer to some of these questions.

Much of the current research on sharks seeks to answer these questions and more about their ecology, the interrelationships between them, and their living and non-living environment. As sharks come under more intense threats to their survival—from overfishing, habitat degradation, climate change, and other human insults—knowing their ecological roles and life history characteristics is critical to developing plans to effectively manage and conserve them and the ecosystems in which they live.

SHARK PREDATOR TYPES

We have already established that most sharks are not apex predators; that is, they are not at the very highest level of the food web. Apex predators are characterized by their large size and the fact that they have no predators as adults—although this is an oversimplification since even White Sharks (*Carcharodon carcharias*; page 106) are occasionally hunted and killed by Orcas (*Orcinus orca*), which then eat only the liver and other organs and leave most of the carcass untouched. On coral reefs, Tiger Sharks (*Galeocerdo cuvier*; page 232), Bull Sharks (*Carcharhinus leucas*; page 196), Great Hammerheads (*Sphyrna mokarran*; page 44), and Silvertip Sharks (*Carcharhinus albimarginatus*) are considered true apex predators. Elsewhere, others include White Sharks, Oceanic Whitetip Sharks (*Carcharhinus longimanus*; page 140), Greenland Sharks (*Somniosus microcephalus*; page 202), and Bluntnose Sixgill Sharks (*Hexanchus griseus*; page 38).

← A Sand Tiger (*Carcharias taurus*), also called Ragged Tooth Shark and Grey Nurse Shark, glides at a deceptively slow speed above a school of fish at Aliwal Shoal, South Africa. In pursuit of prey, this top predator is capable of rapid bursts of speed.

ECOLOGICAL ROLES OF SHARKS

Sharks found along coral reefs

On coral reefs, Tiger Sharks, Bull Sharks, Great Hammerheads, and Silvertip Sharks are considered true apex predators. When these are absent, Grey Reef Sharks and Caribbean Reef Sharks may fill that ecological role.

ECOLOGICAL ROLES OF SHARKS

In mangrove estuaries, Bull Sharks and Smalltooth Sawfish (*Pristis pectinata*; page 112) share the top predator space.

Most sharks on coral reefs are considered mesopredators (page 93). In fact, most sharks—including the Spinner Shark (*Carcharhinus brevipinna*) and Porbeagle (*Lamna nasus*; page 110) are mesopredators. On a five-point scale, with higher numbers associated with higher-level predators, mesopredatory shark trophic levels range from about 3.1 (Zebra Shark, *Stegostoma fasciatum*) to 4.3 (requiem sharks such as the Caribbean Reef Shark, *Carcharhinus perezi*—page 234, and Grey Reef Shark, *C. amblyrhynchos*). Trophic levels for sharks considered apex predators range from 4.3 (Bluntnose Sixgill Shark) to 4.7 (Broadnose Sevengill Shark, *Notorynchus cepedianus*). The trophic level of the White Shark is 4.4. The average for all sharks whose trophic levels have been assessed is greater than 4, which means that most sharks are considered tertiary consumers. For comparison, the trophic level for Orcas is 4.5. Of course, these numbers represent only estimates and are not absolute, but the relative values are corroborated by what is known about the life histories of these sharks.

CHALLENGES AT THE TOP

Apex predators face unique challenges since they begin with a huge disadvantage: there are fewer of them, a fact of life determined in part by the laws of thermodynamics. The classic triangular trophic pyramid graphically depicts the low density of apex predators at the tip. As the pyramid widens, the biomass and number of organisms at each level increases, with the widest levels belonging to the first trophic level, the primary producers (e.g., phytoplankton and seagrass), followed by the second trophic level, the primary consumers (zooplankton and grazers).

← Grace personified, Caribbean Reef Sharks (*Carcharhinus perezi*) patrol the rock and corals near Grand Bahama, The Bahamas.

LIFESTYLE CHOICES

Sharks employ one of two strategies: being large and very active, and eating larger prey; or being sluggish and feeding lower on the food web. Apex predatory sharks such as the White Shark are very active, and their daily ration (defined as the mean percentage of an organism's total weight that is consumed over a 24-hour period) reflects this: about 1.5–1.8. In other words, they must consume a large amount of prey—on average, a 2,200 lb (1,000 kg) White Shark needs to eat about 40 lb (18 kg) of prey a day.

The alternate strategy is employed by the three planktivorous filter-feeding sharks—Whale Shark (*Rhincodon typus*; page 144), Basking Shark (*Cetorhinus maximus*; page 46), and Megamouth (*Megachasma pelagios*; page 108), all of which have trophic levels below 3.4. These species are all slow-moving, thus requiring less food, and they eat lower on the food web, where there is great prey biomass and thus more energy, than apex predators and mesopredators. Although estimating the daily rations for the three species of plankton-eating sharks is logistically difficult due to their extremely large sizes, these figures are likely to be much lower than for a White Shark.

Why does this decrease in density occur? As a rule of thumb, only about 10 percent of the energy in one trophic level is available to create biomass at the next highest trophic level; the remaining 90 percent is either lost as heat (since energy transformations are inefficient, which is the reason why we mention the laws of thermodynamics, above) or is used by the organisms to power their metabolism—for example, to swim, grow, mate, and so on. Moreover, foraging and grazing do not consume all of the energy of lower trophic levels. Thus, to build 2.2 lb (1 kg) of a shark with a trophic level of 4.0, such as the Atlantic Sharpnose Shark (*Rhizoprionodon terraenovae*; page 200), requires 22 lb (10 kg) of its trophic level 3 prey (e.g., small fish and crustaceans). In turn, the small fish and crustaceans of trophic level 3 would need to consume 220 lb (100 kg) of prey at trophic level 2, which themselves would require 2,200 lb (1,000 kg) of primary producers at the first trophic level. The average weight of an adult Atlantic Sharpnose Shark is about 15.4 lb (7 kg), so using our crude estimate, it would take 15,400 lb (7,000 kg) of phytoplankton to build. And the example we chose is a mesopredator, so to build a hypothetical 110 lb (50 kg) apex predatory shark that exclusively consumed Atlantic Sharpnose Sharks would require a staggering 110,000 lb (50,000 kg) of phytoplankton. Apex predatory sharks are thus particularly dependent on their entire ecosystem for their nourishment, not just their prey, and even healthy apex predator populations must exist at very low densities relative to species at lower trophic levels.

↑ A Tiger Shark (*Galeocerdo cuvier*) deploys its white nictitating membranes, which typically indicates that the shark perceives a threat or detects a meal.

← A Whale Shark (*Rhincodon typus*) immersed in azure off the coast of Isla Mujeres, Mexico.

A FINE BALANCE

We know that apex predators consume animals below them in the food web, but their ecological role is more complicated. In some ecosystems, it is thought that apex predators and other organisms near the top of the food web exert top-down regulation of the structure of the biological community of that system. In other words, they play a major role in controlling the diversity and abundance of their prey, as well as their prey's behavior, and have an impact on other organisms at lower trophic levels.

As there are fewer apex predatory sharks to begin with, it is easy to see how the loss of even a small number can lead to negative consequences for their entire ecosystem. Let us revisit coral reefs, ecosystems where overfishing and other stressors have resulted in decreased abundance and even total loss of their top predators. Apex predators on coral reefs include Great Hammerhead, as well as Bull, Tiger, and Silvertip Sharks, along with some bony fishes such as Great Barracudas (*Sphyraena barracuda*). Mesopredators on coral reefs include medium-sized sharks such as Caribbean Reef and Grey Reef Sharks (although on some reefs these two species may be apex predators), as well as smaller species such as the Blacknose Shark (*Carcharhinus acronotus*) and Halmahera Epaulette Shark (*Hemiscyllium halmahera*; page 80) on Caribbean and Indo-Pacific reefs, respectively. Even species that are apex predators as adults may be mesopredators as juveniles.

Theoretically, if top predators are removed from an ecosystem, phenomena called prey (or predation) release and trophic cascade result. The former refers to large population increases of the prey associated with the predator that was removed, and the latter are the reverberations of this action down the food web. Trophic cascades are less likely to occur in biodiverse ecosystems with complex food webs where functional redundancy and trophic overlap ameliorate such effects. For example, in an estuary where there is a suite of many species with overlapping diets that occupy the upper predatory roles, the effects of the loss of one, or even a few, species may be compensated by other predators and may not induce a trophic cascade. But in ecosystems such as coral reefs where, despite the high diversity of primary and secondary consumers (small fishes and invertebrates), the food web is relatively simple and linear and there are defined apex predators, their loss is likely to elicit such cascading effects. Clear and unequivocal evidence of trophic cascades on coral reefs due to removal of top predators is not yet available. What is known is that prey densities of some species have increased when shark populations have been reduced—for example, in northwest Australia after Silvertip and Grey Reef Shark populations dropped on reefs. Another line of evidence consistent with the trophic cascade hypothesis came from a study showing that after sharks were removed from a reef, the diets of their prey shifted from invertebrates to small fish, which apparently had become more numerous. Coral reefs are already under siege from climate change, pollution, and other threats, but removal of top predatory sharks from this ecosystem, even in the absence of irrefutable evidence of resulting trophic cascades, clearly only adds to the damage.

↖ A silhouette of grace and beauty over powdery blue sands, this Great Hammerhead (*Sphyrna mokarran*) swims in the shallows of Bimini, The Bahamas.

Studying sharks

Studying sharks is a rewarding and meaningful activity, but it is fraught with challenges. Not least of these is the fact that sharks live in water, which poses numerous problems for humans. While we have made advances in breathing while underwater, at least to relatively shallow depths, our movement is still awkward and slow, and we are not very good at concealing ourselves. And, of course, most sharks do not live in clear, shallow water.

CATCHING SHARKS FOR RESEARCH

Aside from taking to the water to observe sharks, researchers also capture them to study them. Sharks caught on longlines, hook-and-line, gillnets, and trawls have provided vital information on life history characteristics (diet, age, rate of growth, reproductive information, etc.), effects of pollution, anatomy, and so on. If sharks survive the trauma of being captured, many species—particularly benthic and smaller, shallow-water inhabitants—can be maintained in captivity, during which time aspects of their behavior, physiology, toxicology, and other characteristics can be elucidated.

Catching sharks is not without its problems. The first of these is the high cost of sampling at sea. Second, capture stress can be fatal in many species, belying their reputation of hardiness. Consider the Cuban Dogfish (*Squalus cubensis*), a common bycatch species in deep snapper and grouper fisheries in the Gulf of Mexico. Although about 97 percent of dogfish caught are reported to be released alive, about 50 percent of those die in a day. Great Hammerheads also experience high post-release mortality. Third, many studies require large sample sizes (for example, as many as 100 individuals are required to determine a shark's diet using analysis of gut content). Achieving the minimum sample sizes required for studying depleted populations of apex predators, which have lower populations to begin with, is particularly challenging. Finally, being bitten while handling sharks is a real occupational hazard for shark biologists!

OTHER STUDY METHODS

More recent advances in studying sharks involves advanced deep-diving submersibles (including remotely operated vehicles), camera traps, and environmental DNA. The latter technique, often shortened to eDNA, is becoming increasingly useful for determining the presence of sharks and other species in an ecosystem. It is based on the assumption that every living organism sheds fragments of itself (e.g., skin cells, fecal matter) wherever it has been. Thus, a sample of water can tell you whether a particular species of shark was present, although it cannot tell you when it was there or how many individuals were present.

One widely employed study method, tag-and-release fishing, is used by both citizen scientists (i.e., the public), shark research scientists, and some commercial fishers. A variety of tags can be employed, with the common denominator that they can be applied quickly, such that they will stay in place and not harm the shark. This method is explored in more detail on the following pages.

→ The sonogram of the uterus of the Tiger Shark (top) is displayed during the examination. Information gathered from ultrasonography provides invaluable information on a species' life history characteristics, which are then used in conservation and management decisions. Scientists from the University of Miami (middle) secure a satellite tag to the dorsal fin of a Tiger Shark in The Bahamas in a procedure that takes only minutes. Studies of shark movements using satellite telemetry have provided insight into shark habitat use and migrations. A Tiger Shark (bottom) swims at a provisioning station as divers observe, in Grand Bahama, The Bahamas.

← Scientists secure a Tiger Shark with a rope on the pectoral fin, prior to measuring it, taking tissue samples, and affixing a tag. A slip of the ropes securing it, and the shark swims away into its Grand Bahama environs.

THE ECOLOGY OF SHARKS

Common external shark tags

Surveys that rely on catching sharks with longlines and then tagging them are critical to understanding the health of shark populations and delineating areas that represent vital habitat for sharks or where overfishing is occurring. They can also shed light on life history characteristics such as growth, migrations, and population structure. Two broad categories of tags are used: conventional and telemetry.

Three commonly used tags to track shark movements
M-type tags require recapture of the shark by a cooperative fisher in order to convey information. Satellite tags, in contrast, upload information to orbiting satellites and can provide real-time information on a shark's location. External acoustic tags house ultrasonic transmitters that transmit a ping, and which require a nearby receiver or directional hydrophone attached to a receiver.

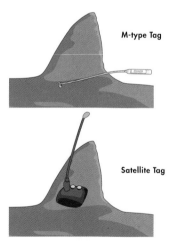

M-type Tag

Satellite Tag

Acoustic Tag

CONVENTIONAL TAGS

Conventional tagging requires that an external tag be anchored to the shark. These tags can be nylon or steel darts that are inserted into the tissue supporting the dorsal fin using a stainless-steel needle, as well as tags similar to those used on the ears of livestock, which are attached to the dorsal fin itself. Disk tags can also be used, attached on each side of the fins or muscle.

In all cases the tag is labeled with a unique identifying number and the researcher's contact information. Thus, the tagged shark must be recaptured or observed by someone who is willing to report the recapture and is capable of doing so. Tags may last the life of the shark—20 or more years, with the record standing at almost 42 years for a School Shark (*Galeorhinus galeus*) recaptured in southern Australia.

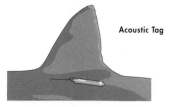

→ Scientists (top) using ultrasound on a mature female Tiger Shark to determine whether it is pregnant and, if so, how far along and how many offspring are inside. The shark is in a calm state known as tonic immobility. A tagged Tiger Shark (bottom) immediately after it was released.

Return rates are normally less than 5 percent. The authors tagged nearly 500 newborn Atlantic Sharpnose Sharks in North Inlet, South Carolina, and were very pleased to get reports of 14 recaptures over one summer, as far as 20 miles (30 km) away. We also tagged nearly 6,000 juvenile Sandbar Sharks (*Cacharhinus plumbeus*; page 198) in Chesapeake Bay, Virginia, some of which were recaptured more than 20 years later and as far away as Port Aransas, Texas, some 2,000 miles (3,200 km) from the original tagging site.

Another form of conventional tagging involves the use of passive integrated transponders (PIT tags), rice-grain-sized microchips inserted into muscle tissue of sharks. The PIT tag essentially gives the shark a lifelong barcode, like those on products at a store that are scanned at the register for the price. The technology is similar to the way automatic toll roads detect your vehicle and charge your account. Upon recapture of a PIT-tagged shark, an electric reader uses an electromagnetic field to activate the PIT tag, triggering it to transmit its identification number.

TELEMETRY TAGS

Telemetry tags can be radio, acoustic, or satellite. Radio signals do not transmit well in salt water, so for sharks, these are used only in species that inhabit freshwater rivers. In acoustic telemetry, transmitters are attached externally to the shark, much like conventional tags, or they are surgically implanted in sharks in a procedure that takes just a few minutes. Acoustic transmitters are tracked either actively or passively. In active tracking, the transmitter emits a series of pings every second or less. The researcher uses a receiver and headphones to actively listen for the shark and follow it around, recording its location every few minutes, thus providing detailed habitat information. In passive telemetry, each transmitter randomly broadcasts a unique coded signal in intervals—typically from 40 seconds to 180 seconds—for as long as the battery lasts, which may be 10 years in some cases. A receiver capable of detecting the transmitter's pings must be present and within 110–880 yd (100–800 m) of the shark or it does

↗ Fisheries science researchers restrain an adult Sandbar Shark (*Carcharhinus plumbeus*) in preparation for measuring and tagging it prior to release.

→ A Silky Shark (*Carcharhinus falciformis*) carries a pop-off archival satellite tag (PSAT) anchored into the muscle. The tag will record temperature, depth, and light data (used to estimate location) until releasing from the shark and uploading the data through a satellite link. A streamer-style dart tag is also attached beneath the first dorsal fin.

not detect the animal's presence. Fortunately, arrays of receivers have been deployed by cooperating groups of researchers across wide geographic areas, and these researchers share data detected on their receivers with the scientists who implanted the acoustic tags.

The most sophisticated and expensive telemetry tags (US$1,000–5,000 apiece) are satellite (SAT) tags. SAT tags transmit radio signals when the tag's antenna is above the water, and these are picked up by orbiting satellites. One common type of SAT tag is the smart position and temperature (SPOT) transmitting tag, which is typically attached to the dorsal fin of the shark or towed behind it with a tether. SPOT tags can provide real-time movement data as long as the shark is near the water's surface while satellites are passing overhead. Archival SAT tags store light, time, depth, and temperature data onboard. They are programmed to detach from the shark at a specific time, float to the surface, and transmit the data to satellites. Light data allow calculation of day length (determines latitude) and timing of solar noon (indicates longitude), providing location estimates of the tagged shark. SAT tags can provide a plethora of information, such as the depth and temperature of the environment where the shark has been, its swimming speed, and aspects of the shark's physiology and behavior. As newer technologies emerge, scientists will be able to develop even more sophisticated and smaller tags, capable of measuring behavioral and physiological aspects of the biology of sharks and other species.

THE ECOLOGY OF SHARKS

WHAT HAVE SCIENTISTS LEARNED FROM TAGGING SHARKS?

Tagging studies have provided information on the movements and migrations of sharks, and their habitat preferences, all of which are vital to understanding their ecology. Data on where a shark feeds, mates, and gives birth or lays its eggs can be used to protect habitats most susceptible to human degradation or where sharks are exposed to fisheries capture.

Tracks of eight satellite-tagged Tiger Sharks in the northwest Atlantic

The tags were applied in Bimini, The Bahamas, by the Bimini Biological Field Station as part of a study to gain insight into movement patterns and residency times of this large predatory shark. Note the variability in movement patterns among individual sharks. The colors represent individual sharks tagged at different times.

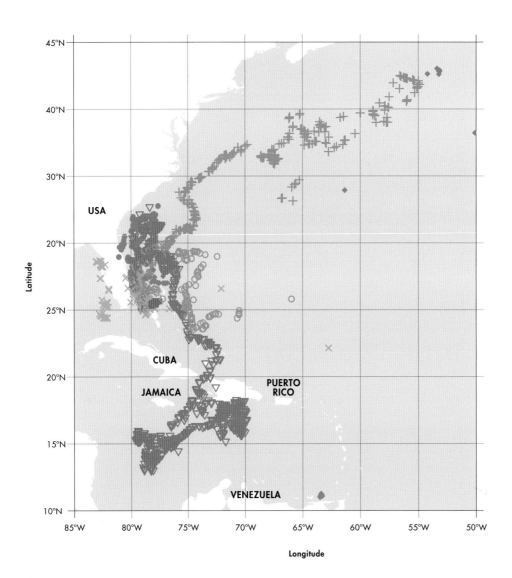

Many sharks use specific mating and nursery areas (where sharks or born or where juveniles spend a disproportionate amount of time), and tagging studies can delineate these. For example, a study using conventional tagging found that Chesapeake Bay was the largest summer nursery for Sandbar Sharks in the western Atlantic. Moreover, the study showed that younger Sandbar Sharks (below the age of three years) moved to North Carolina in winter, then moved back into Chesapeake Bay each summer, whereas older sharks expanded their range further south in winter and used waters outside of Chesapeake Bay in summer. These data were used by the state government to enact restrictions on gillnetting in habitats used by the juveniles to protect this vital age group.

Tagging studies inform scientists about the home range of sharks, the area they use routinely for their daily activities. Researchers at the Bimini Biological Field Station in The Bahamas have been PIT-tagging juvenile Lemon Sharks (*Negaprion brevirostris*; page 230) for nearly three decades, and have found that they establish small, well-defined home ranges in shallow water within their nurseries. Newborn Blacktip Reef Sharks (*Carcharhinus melanopterus*) around Moorea, French Polynesia, also have small home ranges, estimated to cover an area of just 1,500 sq ft (140 sq m). The sharks in this study lived adjacent to a deep channel, which constituted a barrier to them. Benthic sharks typically also have smaller home ranges. If you observe these species while you are snorkeling, the chances are that there are more in the area. At the other extreme are more nomadic sharks that range widely within their habitat, including Blue Sharks (*Prionace glauca*; page 138) and Shortfin Makos (*Isurus oxyrinchus*; page 74). In 2019, researchers were able to attach a satellite tag to a Bluntnose Sixgill Shark in The Bahamas from a submersible at a depth of 1,700 ft (520 m) using a remotely controlled speargun. Although the species has been tagged at the surface after capture on a deep longline, this groundbreaking tagging episode was the first for the species in a more natural setting, involving less stress on the shark. Previous tagging studies for this species uncovered a diel (over a 24-hour period) vertical movement cycle, shedding light on the habitat preferences of this ancient species. In Hawai'i, researchers found that the sharks ascended to the depth of the oxygen minimum zone (1,600–2,600 ft/500–800 m) during daytime.

↗ A Bluntnose Sixgill Shark (*Hexanchus griseus*). At lengths of over 18 ft (5.5 m), this species (along with sleeper sharks) breaks the rule that deep sea sharks are small.

THE ECOLOGY OF SHARKS

How long do sharks live and how fast do they grow?

Knowing where sharks live and move to is critical, but this information is just one part of their ecological story. As a group, sharks are slow-growing and long-lived. But how slow? How long? How is this information determined? More specifically, for every species shark ecologists seek to determine the size and age at maturity, the size and age at first reproduction, the fecundity, the frequency of reproduction, the survival of offspring, and the maximum size and age.

Let us begin with age and growth. Like tree rings, hard parts in the bodies of sharks record their age. These include the vertebral centra, the main structural elements of the backbone and, in some species, spines on the dorsal fins. Studies into these have shown that sharks are long-lived—for example, White Shark (*Carcharodon carcharias*; page 106) age estimates range from 33 years to more than 70 years, and estimates of the maximum age of the Greenland Shark (*Somniosus microcephalus*; page 202) extend to at least 272 years!

Sharks are also slow-growing, with comparatively late ages of maturity. Late maturity typically results in larger offspring, each of which has a higher likelihood of survival. There is, however, a trade-off. To produce larger young requires higher maternal investments and results in lower fecundity. Fewer offspring are also associated with small body size in most shark species, unless those offspring develop external to the mother (i.e., egg-laying species).

↘ Timeline depicting some major events that would have occurred in the long life of a 250-year-old Greenland Shark (*Somniosus microcephalus*).

→ A Greenland Shark swims in the frigid waters under the ice in the Canadian Arctic. The filament attached to its eye is a parastic copepod.

1750	1776		1784	1787	1789	
[0yrs]	[27yrs]		[34yrs]	[37yrs]	[39yrs]	
	United States Declaration of Independence is adopted by the Second Continental Congress		End of Enlightenment	Constitution of the United States is composed at the Constitutional Convention	French Revolution	

102

1835
[85yrs]
Charles Darwin visits the Galapagos Islands

1865
[115yrs]
Civil War ends in the United States

1918
[168yrs]
World War I ends

1945
[195yrs]
World War II ends

2000
[250yrs]
Start of the 21st century

THE ECOLOGY OF SHARKS

AGE OF MATURITY COMPARISONS

That *trophy* 6.5 ft-long (2 m) Tiger Shark (*Galeocerdo cuvier*; page 232) on the wall of your favorite restaurant may be only two years into its life, and six or seven years away from sexual maturity at about 10 ft (3 m). Sandbar Sharks in Chesapeake Bay are born at a length of about 16–20 in (45–50 cm) and females reach sexual maturity after around 15 years at about 6 ft (1.8 m). Male and female White Sharks mature, respectively, at around 26 years (about 13 ft/4 m) and 33 years (about 15 ft/4.5 m). At the other end of the scale, Bonnetheads (*Sphyrna tiburo*; page 204) mature at three years.

↑ A Sandbar Shark off Hawai'i, where they occupy deeper water than in the Atlantic Ocean.

→ Small sharks, including a mix of Brazilian Sharpnose Sharks (*Rhizoprionodon lalandii*), Caribbean Sharpnose Sharks (*R. porosus*), and juvenile Scalloped Hammerheads (*Sphyrna lewini*), are destined to become "Bake and Shark," a popular traditional dish in Trinidad.

HOW LONG DO SHARKS LIVE AND HOW FAST DO THEY GROW?

Age of maturity can also vary geographically. Maturity of Sandbar Sharks occurs at about 16, 15, and 10 years, respectively, for populations in the Indian Ocean, northwest Atlantic Ocean, and the Pacific Ocean off Hawai'i. The live-bearing deep-sea sharks (including gulper sharks, dogfishes, sleeper sharks, and some lantern sharks) have extremely conservative life histories, reaching maturity as late as 35 years in some species.

In keeping with the slow life history characteristics of sharks, gestation periods can be as short as five months (Bonnethead), to as long as two years (Spiny Dogfish, *Squalus acanthias*; page 264) or even more than three years (Frilled Shark, *Chlamydoselachus anguineus*; page 168).

The number of offspring is also variable. The mean fecundity of cat sharks, which are egg-layers and are small (less than 3 ft/1 m), is 60 eggs. Dogfish, which give birth to live young and are much larger on average, average only 4.6 pups. At least one gulper shark species gives birth to just a single pup. Although they mature relatively early (9–10 years for females), Sand Tigers (*Carcharias taurus*; page 236) produce only two offspring, each about 3 ft (1 m) long. Whale Sharks (*Rhincodon typus*; page 144) give birth to as many as 300 pups.

KNOWLEDGE GAPS

As this chapter has demonstrated, the ecological roles sharks play in marine ecosystems are far more complex than the simplistic notion that they are all apex predators. While sharks play critical roles as planktivores, mesopredators, and apex predators in their ecosystems, these roles are not well known in many cases. Huge gaps in our knowledge of the ecology of sharks exist for entire ecosystems—for example, the deep sea, where even the number of shark species is a mystery. Research on life history characteristics of deep-sea and other sharks using traditional methods such as longlines and newer technologies such as telemetry tagging is still needed.

Educating the public about the ecological importance of sharks and the threats to their health is also a must. The growing global human population is placing increasing demands on marine ecosystems to supply food, while at the same time threatening the existence of entire taxa by warming the planet, acidifying the oceans, overfishing, destroying habitats, and otherwise polluting the marine environment. For marine ecosystems to function—that is, to possess the biodiversity necessary for all trophic levels to prosper for the long term without humans irreversibly degrading these ecosystems—they require the appropriate species and populations of sharks.

THE ECOLOGY OF SHARKS

CARCHARODON CARCHARIAS

White Shark

Power, grace, and *Jaws*

SCIENTIFIC NAME	*Carcharodon carcharias*
FAMILY	Lamnidae
NOTABLE FEATURE	Long, conical snout; stout body; long gill slits; keel on caudal peduncle; crescent-shaped (lunate) caudal fin; serrated, triangular teeth
LENGTH	20 ft (6 m)
TROPHIC LEVEL	Apex predator of marine mammals, sharks, and other large fish

The White Shark is the most widely recognized shark species, in part due to its starring role in the 1975 movie *Jaws*. Despite this, however, and the fact that White Sharks are among the most studied, scientists still do not know the species' ecology well. Shark biologists have not observed White Sharks mating or pupping, nor are the specific habitats vital to neonates and juveniles known. Catch records from commercial and recreational fishers in southern California of small individual White Sharks, some with yolk-sac scars, suggest that this area might be a nursery for the species, along with the west coast of the central Baja California Peninsula, Mexico, and the New York Bight on the US East Coast. Knowing life history characteristics such as these is critical to managing the species and the ecosystems in which it lives.

White Sharks are globally distributed in both coastal and oceanic waters of 54–75°F (12–24°C). They are the largest of all living top predatory fishes—a full-grown, 20 ft (6 m) White Shark can weigh an impressive 4,200 lb (1,900 kg).

Extraordinary as this may seem, an adult White Shark would be dwarfed by the largest predatory fish ever to swim the world's oceans, the extinct Megalodon (meaning "giant tooth"), whose maximum size may have been 60 ft (18 m). Contrary to public perception, Megalodon was not an ancestor of the White Shark but rather an evolutionary dead end, and an offshoot of the closely related mako lineage. White Sharks migrate latitudinally as well as longitudinally across oceans. A 12.5 ft (3.8 m) female tagged off South Africa was detected 99 days later off Australia, more than 6,800 miles (11,000 km) distant, before returning to South Africa.

As predators, White Sharks have few equals in the marine realm, in large part because of their ability to retain heat—as much as 26 °F (14.3 °C) above ambient temperatures—which translates into increased muscle power and sensory acumen. White Sharks are known to breach when attacking seals from beneath, hurling their body meters into the air. That said, Orcas (*Orcinus orca*) have been known to kill White Sharks, off both California and South Africa, and eat their livers.

White Sharks (and their close relatives, including the other mackerel sharks) use oophagy for embryonic nourishment. In this strategy, which means "egg-eating," the embryos initially use their yolk sac for energy. After the yolk is absorbed, the mother begins to ovulate unfertilized ova. The White Shark embryos then swim within the uteri, eating up these ova and storing them in their large egg stomachs.

→ Mouth agape in a pose depicting its predatory prowess, a White Shark breaches in pursuit of a seal decoy in False Bay, South Africa.

THE ECOLOGY OF SHARKS

MEGACHASMA PELAGIOS

Megamouth
Plankton eater lurking in the deep

SCIENTIFIC NAME	*Megachasma pelagios*
FAMILY	Megachasmidae
NOTABLE FEATURE	Large head with fleshy lips, rounded snout, terminal mouth
LENGTH	18 ft (5.5 m)
TROPHIC LEVEL	Planktivore, primarily feeding on euphausiid shrimp (krill)

The Megamouth shark was not even known to science until 1976, when a male specimen was snagged by one of two parachutes deployed to depths of 540 ft (165 m) as sea anchors by a US Navy oceanographic research vessel near Hawai'i. The specimen was subsequently frozen, after which it was described and named *Megachasma pelagios*, loosely meaning "great gape of the pelagic," after its most prominent feature, to which its common name also refers.

Imagine the excitement of shark scientists at the discovery of such a huge beast of a shark (14.6 ft/4.4 m in length and weighing 1,650 lb/750 kg), distinct enough to merit being in its own family. The next specimens were not caught until 1984 and 1987. Reports of the capture of additional Megamouth sharks trickled in over subsequent decades; by the end of 2006, 38 had been captured or observed.

How had such as enormous animal evaded discovery for so long? As it turns out, it had not. Taiwanese commercial drift-net fishers had frequently caught the species as bycatch, but the scientific community had been unaware of these captures. Certainly, other fishers have also caught specimens, but perhaps did not notice the significance of their catch. To date, there have been 269 confirmed captures or sightings of this shark, over half of them from Taiwan. That said, the species is still poorly known, as might be expected for such a rarely encountered animal. The Megamouth is believed to be the most primitive member of the living lamniform sharks.

The species likely takes part in the largest animal migration on the planet, the daily vertical migration of what is called the deep scattering layer. This large assemblage comprises fish, squid, and crustaceans that live at depths of 980–1,640 ft (300–500 m) or even deeper during the day, but migrate to shallow water in the evening. The Megamouth shark is a planktivore, mostly feeding on krill at night. However, instead of straining the water as it flows through the mouth and over gill rakers front to back like other planktivores, it swallows water and then forces it forward through the gill rakers, which then strain out the plankton.

→ A Megamouth off California. Hotspots for the species, which is thought to be a seasonal, latitudinal migrator, include Japan, Taiwan, and the Philippines. Most aspects of their biology remain a mystery.

THE ECOLOGY OF SHARKS

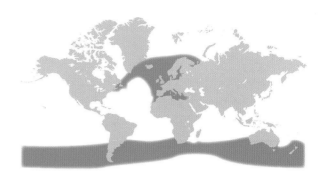

LAMNA NASUS

Porbeagle

Overfished, warm-blooded mesopredator

SCIENTIFIC NAME	*Lamna nasus*
FAMILY	Lamnidae
NOTABLE FEATURE	Conical snout, streamlined body, two caudal fin keels, rear tip of dorsal fin white
LENGTH	11.5 ft (3.5 m)
TROPHIC LEVEL	Opportunistic mesopredator of fish and squid

As one of the five shark species in the family Lamnidae, the Porbeagle (note, as with its cousin the Shortfin Mako, the word "shark" is not included in its common name) is warm-blooded—or, to use the precise physiological term, endothermic. Endothermy endows this species with the same set of benefits and costs as its close relatives. The former include a high-performance metabolic system, more acute senses, and the ability to be less restricted by low environmental temperatures when it dives or migrates seasonally into cooler waters. The cost of endothermy is that the metabolic machine, especially the powerful muscles, must be fueled—in other words, the diet of Porbeagles must include more calories than those of less active sharks.

Unlike the Porbeagle's close relatives, endothermy did not lead it or its North Pacific Ocean ecological counterpart, the Salmon Shark (*Lamna ditropis*; page 136), in the direction of being an apex predator. Instead, the Porbeagle feasts on squid and a variety of small to medium-sized bony fish such as mackerel, herring, and hake.

The species' story is one of tragic overfishing. The meat of Porbeagles, like that of Shortfin Makos, is quite tasty and has a firm texture. Norwegian longline fishers began to fish commercially for Porbeagles in 1961, and landings for that year were 2,000 tons (1,800 tonnes). At the peak of fishing in 1964, landings had reached 10,300 tons (9,300 tonnes). In 1967, the fishery collapsed, and by 1970 the total catch was around 1,100 tons (1,000 tonnes). Some commercial fishing for Porbeagle is still conducted by Canada and the Faroe Islands. While overfishing no longer occurs, the population in the northwest Atlantic has still not fully recovered. As a result, the IUCN considers the Porbeagle Critically Endangered in the northeast Atlantic Ocean and Mediterranean Sea, Endangered in the northwest Atlantic, and Vulnerable globally.

→ A Porbeagle, the ecological counterpart to the Salmon Shark. The common name Porbeagle is said to originate from Cornish and recalls the beagle dog, in reference to the propensity of the species to hunt in "packs."

THE ECOLOGY OF SHARKS

PRISTIS PECTINATA

Smalltooth Sawfish

Critically Endangered shark-like ray

SCIENTIFIC NAME	*Pristis pectinata*
FAMILY	Pristidae
NOTABLE FEATURE	Shark-like body with elongated, toothed snout (saw)
LENGTH	16 ft (5 m)
TROPHIC LEVEL	Upper-level predator of bony fish, stingrays, shrimp, and crabs. In mangrove estuaries, it surprisingly shares the top predator space with Bull Sharks

Despite being a ray, the Smalltooth Sawfish is included here in a book about sharks because, first, all batoids (rays and skates) share a common heritage and most distinguishing features with sharks, having diverged from the shark lineage about 270 million years ago. In fact, rays are sometimes called "flat sharks" (not to be outdone, batoid biologists call sharks "sausage rays"). Second, the Smalltooth Sawfish looks shark-like or, to be more precise, its derived body form has converged with that of sharks. Finally, its story of loss and survival as a species is compelling, cautionary, and perhaps hopeful.

The feature that reveals the Smalltooth Sawfish to be a batoid and not a shark is the location of its gill slits. In sharks, the gill slits are always located above, or dorsal to, the well-defined pectoral fins. In batoids, they are located on the underside of the body, below where the broadly expanded pectoral fins—which we call the wings in most rays—are attached to the body. The saw of this species and its four sawfish cousins is actually an elongated rostral cartilage equipped with sharp, flattened teeth. The rostrum, which houses the electrosensory ampullae of Lorenzini, is used to impale and stun fish.

This Critically Endangered species is found in tropical and subtropical coastal areas on both sides of the North Atlantic, but has been extirpated from most areas, especially in the east. It was placed on the US endangered species list in 2003. Continued threats include coastal development, dredging, mangrove removal, seawall construction, alteration of freshwater flow, habitat fragmentation, climate change, and especially commercial fishing as bycatch. There is a glimmer of hope, however: extensive efforts to reverse population declines in southwest Florida are beginning to yield positive results.

→ The endangered Smalltooth Sawfish in Everglades National Park, Florida. In 2017, the present author Dean Grubbs played obstetrician when a large female Smalltooth Sawfish caught in The Bahamas went into labor while blood was being collected and a transmitter implanted. Five pups were successfully delivered, and all swam strongly away.

THE ECOLOGY OF SHARKS

TRIAENODON OBESUS

Whitetip Reef Shark

Nocturnal predator on coral reefs

SCIENTIFIC NAME	*Triaenodon obesus*
FAMILY	Carcharhinidae
NOTABLE FEATURE	Slender shark, conspicuous white tips on first dorsal fin and upper lobe of caudal fin
LENGTH	6.9 ft (2.1 m)
TROPHIC LEVEL	Nocturnal mesopredator of fish and invertebrates on coral reefs

Not to be confused with the epipelagic and much larger Oceanic Whitetip Shark (*Carcharhinus longimanus*; page 140), the Whitetip Reef Shark is found in shallow water in caves and on coral reefs in the Indo-Pacific. It forages at night and rests during the day. While resting, it pumps water over its gills, a feat that most members of its family cannot duplicate. The Whitetip Reef Shark is among the most common shark species on Indo-Pacific coral reefs, and is included here because, in contrast to many of the other sharks in this book, it is relatively well known, especially in terms of its behavior.

Whitetip Reef Sharks are among the most flexible of sharks. This flexibility, in combination with their slender body, enables them to navigate the tight curves and sharp projections of the coral reefs they inhabit. It also allows them to wriggle into tight crevices as they forage for fish and invertebrates—primarily crustaceans and octopuses—on the reefs. They have relatively small home ranges; in other words, they do not wander far from their home reef.

Their mating behavior is among the best known for any shark. Multiple males—up to five have been observed—pursue and attempt to mate with females. Since female Whitetip Reef Sharks are often larger than males, this group method might increase the chances of successful mating, particularly as larger females can probably outmuscle and deter a single male. One or even two males will bite the pectoral fins of a female, which slows them such that the male–female pair or threesome will sink to the bottom. There, one male will insert and lock one of its claspers into the cloaca of the female and, if the female does not dislodge the clasper, will fertilize the female with a pressurized cocktail of semen and seawater. Gestation lasts five or more months, with litters of 1–5.

→ A Whitetip Reef Shark and a moray eel (*Gymnothorax* sp.) in Roca Partida in Mexico. In some locations, Grey Reef Sharks (*Carcharhinus amblyrhynchos*) gather as Whitetip Reef Sharks forage, and the former prey upon fish displaced from the reef by the latter.

SHARKS OF THE
OPEN OCEAN

Predators of the open ocean

Along with the deep sea (see the next chapter), the open-ocean environment is home to sharks that we are unlikely to encounter except in photographs or videos. This is despite the fact that more than 99 percent of the Earth's water is found in these zones and they cover about 45 percent of the surface of the planet. So if you find yourself floating in the middle of the ocean, far from land, what are the chances you will encounter and possibly be eaten by a shark?

← Pre-contemporary artwork often depicts sharks as demonic beings engaged in a frenzy of carnage. In this painting, survivors from the schooner *Tahitienne*, a victim of a Pacific Ocean storm, are attacked by sharks.

→ An Oceanic Whitetip Shark (*Carcharhinus longimanus*), its vivid countershading evident, investigates the camera and its attached diver. Although capable of making dives to 3,550 ft (1,080 m), these apex predatory sharks spend most of their time at a depth of up to 660 ft (200 m) deep.

WHAT IS THE OPEN OCEAN?

The open-ocean ecosystem is that part of the ocean shallower than 660 ft (200 m), known as the epipelagic zone. It is also called the photic zone, since it marks the limit of sufficient light penetration for photosynthesis. The ocean deeper than 660 ft (200 m) is considered the deep sea, which is a distinct ecosystem, even though many of the sharks we discuss in this chapter—for example, the Shortfin Mako (*Isurus oxyrinchus*; page 74), Oceanic Whitetip Shark (*Carcharhinus longimanus*; page 140), and Crocodile Shark (*Pseudocarcharias kamoharai*)—travel between it and the open ocean to forage or rest.

PREDATORS OF THE OPEN OCEAN

IT WORKS BOTH WAYS

Epipelagic sharks may descend into the deep sea, but the vertical migration can move in the opposite direction as well. Many deep-sea sharks, including the Bluntnose Sixgill Shark (*Hexanchus griseus*; page 38) and Cookiecutter Shark (*Isistius brasiliensis*; page 164), venture into the epipelagic zone.

SHARKS OF THE OPEN OCEAN

The open-ocean ecosystem also excludes waters over continental shelves, an area called the neritic zone. Again, many shark species we include in this chapter, such as the Common Thresher (*Alopias vulpinus*; page 142), will penetrate neritic waters, and some neritic, or coastal, species—including the Tiger Shark (*Galeocerdo cuvier*; page 232)—will move offshore. For some species—for example, the Whale Shark (*Rhincodon typus*; page 144)—the lines are even more blurred; they are variously considered as neritic and oceanic (and we include them in both categories). Sharks that do not live on or near the ocean bottom are called pelagic, hence all sharks of the open ocean are also pelagic.

> **IT'S ALL IN THE NAME**
>
> Open-ocean sharks spend the majority of their lives in waters shallower than 660 ft (200 m) and do not predominantly occur on the continental shelf. They can be referred to as open-ocean, oceanic, or epipelagic sharks.

COMMON OCEANIC SHARKS

While more than half of the 543 shark species occupy the deep sea, less than 5 percent of these sharks, or about 25 species, are considered oceanic (although note that this figure is far from definitive). These sharks include many of the familiar larger species, such as the Shortfin Mako and Common Thresher, plus a few that are relatively unknown, including the Longfin Mako (*Isurus paucus*) and Bigeye Thresher (*Alopias superciliosus*). They are all wide-ranging, as you might expect, and many are also distributed around the globe.

This estimate of the number of oceanic shark species is probably liberal, since some of these spend roughly equal time above and below the 660 ft (200 m) depth mark we use to delineate oceanic species. This is especially true for diurnal vertical migrators—that is, those sharks that participate in the world's largest migration, from the deep sea to the epipelagic zone at night, and back to the deep sea during the day. Most of the vertically migrating species barely break into the epipelagic zone during their nighttime excursions to feed at the top of the permanent thermocline. However, some—including the Cookiecutter Shark—can be included as oceanic species because they migrate from the deep sea all the way to the ocean surface each night.

How is this low diversity of oceanic sharks explained? In general, where a shark lives reflects a variety of factors, including the species' range of environmental tolerances and preferences (temperature, salinity, and dissolved oxygen, for example), as well as ecological constraints (including food availability, competition for resources, and risk of predation). The open-ocean ecosystem as a whole is not as productive as coastal ecosystems, which translates into less biomass and a lower biodiversity of large predators, including sharks.

The open ocean is also expansive, with little habitat diversity, few barriers to movement, and fewer ecological niches to fill, so new species do not arise through isolation or specialization, as occurs in coastal and deep-sea habitats. This lack of barriers and isolation is reflected in the population structure of oceanic sharks.

← A Pelagic Thresher (*Alopias pelagicus*) in the Philippines. The jaws of the threshers are very short relative to their width, which is consistent with their diets of small schooling fishes.

A selection of the most common sharks of the open ocean

Foraging preferences for these sharks (drawn roughly to scale) vary by species, but most feed on a wide variety of prey—that is, they are more generalists than specialists.

Shortfin Mako

Blue Shark

Oceanic Whitetip Shark

Common Thresher

Cookiecutter Shark

Silky Shark

Salmon Shark

Porbeagle

Whereas coastal shark species often have many populations across their range, oceanic sharks tend to exist in a single population for an entire ocean basin. For example, there are at least three populations of Blacktip Sharks (*Carcharhinus limbatus*) just along the Atlantic and Gulf of Mexico coasts of the United States, but there is a single Blue Shark (*Prionace glauca*; page 138) population across the entire North Atlantic Ocean, an area bordered by more than 50 countries. This means that oceanic sharks are exposed to fisheries from many countries, and as we discuss below, many species are apex predators. Thus, in addition to low diversity, there are fewer of them, which makes them even more vulnerable to anthropogenic threats.

↓ A Whale Shark (*Rhincodon typus*) feeding among plastic debris. Large debris can clog or perforate digestive tracts or entangle sharks. This debris also degrades into microplastics in the environment, and can enter the food web, often carrying toxic organic chemicals attached to the microplastics.

ON THE MOVE

Oceanic shark species are found primarily between 70°N and 50°S, but they are highly migratory within these latitudes. These migrations can cover great distances, both latitudinally and longitudinally. Several species—for example, Silky Sharks (*Carcharhinus falciformis*; page 272) and Porbeagles (*Lamna nasus*; page 110)—undergo migrations to the equator in winter and back to higher latitudes in the summer. Blue Sharks have been shown to migrate from the southwestern to the southeastern Atlantic Ocean, and in the Pacific Ocean, Salmon Sharks (*Lamna ditropis*; page 136) migrate distances of more than 3,100 miles (5,000 km) and Shortfin Makos distances of 3,400 miles (5,500 km).

Explanations for migrations of open-ocean sharks are not definitive, but they likely are tied to temperature preferences, prey availability, and reproduction. Some individual migrations are astonishing—for example, a White Shark (*Carcharodon carcharias*; page 106) was tracked moving from South Africa to Australia, while a Basking Shark (*Cetorhinus maximus*; page 46) migrated from waters off Massachusetts to Brazil!

GREEN BEANS FOR BLUE SHARKS

Many open-ocean sharks are apex predators, with diets that include bony fishes and squid, and even other sharks and marine mammals. But they are also opportunistic. We once found large quantities of green beans in the guts of three Blue Sharks, presumably discards from passing ships, and you have probably heard of epipelagic sharks found to contain suits of armor, cans of paint, bottles of wine, and equally strange items in their digestive tract. These dietary selections reflect a reality of life in a relatively food-poor environment. Would you pass up a free meal if you weren't certain when another opportunity to eat would present itself, even if it tasted, well, metallic?

PREDATORS OF THE OPEN OCEAN

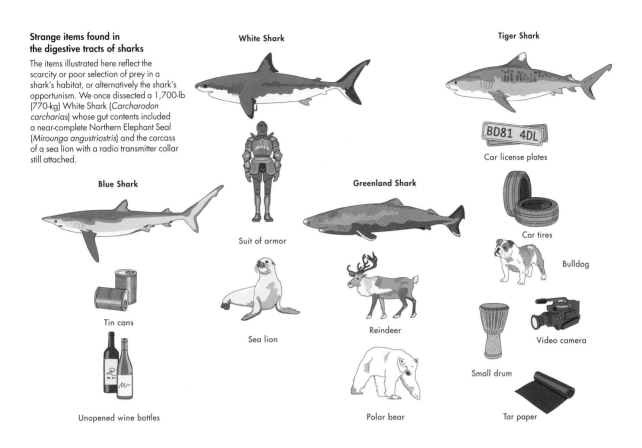

Strange items found in the digestive tracts of sharks

The items illustrated here reflect the scarcity or poor selection of prey in a shark's habitat, or alternatively the shark's opportunism. We once dissected a 1,700-lb (770-kg) White Shark (*Carcharodon carcharias*) whose gut contents included a near-complete Northern Elephant Seal (*Mirounga angustirostris*) and the carcass of a sea lion with a radio transmitter collar still attached.

FOOD FOR THOUGHT

Did you know that White Sharks have their own café in the Pacific Ocean? Located midway between Hawai'i and the Baja Peninsula of Mexico, this zone is about the size of New Zealand and is called the café because early in the twenty-first century scientists discovered that White Sharks migrate to the area and dawdle here. The White Sharks in the area remain in the epipelagic zone during daylight and dive as deep as 660–1,500 ft (200–450 m) at night. Hypotheses conjecture that the sharks in this aggregation were either foraging or mating. The presence of unexpectedly dense concentrations of prey and other organisms at the predators' diving depths led shark scientists to conclude that the White Sharks were at the café to forage. Migrations that give sharks access to prey resources are called alimentary, or trophic, migrations, and are not uncommon.

All three species of planktivorous shark—the Whale Shark, Basking Shark, and Megamouth (*Megachasma pelagios*; page 108)—are considered epipelagic. Their large size reflects the fact that they eat low down on the food web, where prey are most abundant.

Finally, consider the threshers, all of which possess an expanded upper caudal fin lobe that they use to herd and stun the small schooling fishes (such as anchovies) that constitute their major prey (along with squid). The jaws of thresher sharks are very small and weakly calcified, reflecting their small, soft-bodied prey. There is a cost involved in the evolution of this expanded upper caudal lobe. It generates lift and thrust diagonally upward, which would push the anterior body downward in the absence of the concomitant evolution of the threshers' expanded pectoral fins, which put the brakes on this downward force.

SHARKS OF THE OPEN OCEAN

Life history characteristics

As has already been established, the diversity and biomass of open-ocean sharks are both relatively low, a function of their position at the top of the food web and the reduced prey abundance in the open-ocean ecosystem. But are other characteristics of sharks of the open ocean also different from those of sharks in other ecosystems?

LIFE HISTORY CHARACTERISTICS

BODY DESIGNS

Let us first consider body type. Sharks can be assigned to one of several body types depending on the scheme used. Although the overall diversity of open-ocean sharks is low, their morphological diversity is high, reflecting their different lifestyles and predatory strategies.

The mackerel sharks—which include the White Shark, Shortfin Mako, Longfin Mako, Porbeagle, and Salmon Shark, all of which are considered high-performance super-predators—have a streamlined body honed through evolution to be adapted to swimming at high speed. They have a more or less conical head, a deep body, and a crescent-shaped caudal fin. The latter produces increased power and efficiency at high swimming speeds for sustained periods since the shark chases prey capable of achieving high speed themselves, such as tuna.

The mackerel sharks' superior prey-finding ability and powerful musculature, along with their concomitant high swimming speeds, are dependent on their capacity to retain body heat—an extraordinary characteristic for a shark and one that is shared with only a handful of other fishes (most notably, the tunas). A body that is warmer than its environment enables faster chemical reactions within the animal, which underlie all physiological processes and translate into improved senses and more powerful muscle contractions. The cost—and it is not insignificant in a food-poor environment like the open sea—is that the metabolic beast must be fed. If you gamble, don't bet against the mackerel sharks locating and consuming their prey!

← A Shortfin Mako (*Isurus oxyrinchus*), mouth agape, swims in the epipelagic zone of the open ocean. Swimming in these high performance predators, and four closely-related mackerel sharks, is similar to that of the tunas. Like a wind-up fish toy, only the caudal fin and peduncle (narrow posterior section) oscillate, and there is little head yaw (side-to-side motion).

↗ The elongated upper lobe of the caudal fin of the Pelagic Thresher is adapted to herd and stun prey, but it also provides thrust that lifts the rear of the shark during swimming. No worries, though, since the greatly expanded pectoral fins resist being pushed down and also provide lift when the shark moves, thus balancing the shark.

SHARKS OF THE OPEN OCEAN

↑ A Blacktip Shark (*Carcharhinus limbatus*) swims with its tag-along posse of sharksuckers. This widely distributed species is a streamlined, swift-moving, bite-first-ask-questions-later fish-eater, a reference to the mistaken identity bite-and-release interactions with bathers in the murky waters of the southeast US.

↗ A White Shark near Guadalupe Island, Mexico, a small volcanic island and biosphere reserve where White Sharks aggregate. In 2019, about 2,800 tourists cage-dived there. In what ways might this affect the sharks?

LIFE HISTORY CHARACTERISTICS

LIFE HISTORY CHARACTERISTICS

The other large oceanic sharks—including the Blue Shark, Oceanic Whitetip Shark, Silky Shark, and the three species of threshers—are also highly mobile oceanic cruisers, but are not in the same league as the mackerel sharks when it comes to adaptations for high speed. Thus, while their bodies are streamlined, they are less so than in any of the mackerel sharks, and their tails are more typically asymmetrical, which translates into less thrust and acceleration. In addition, their pectoral fins are typically enlarged. These species are capable of short bursts of speed, but the smaller fish and squid on which they feed are not so fast that higher performance is required of the sharks in this category.

Moving swiftly is not a priority for the plankton-eating open-ocean sharks, namely the Whale Shark, Basking Shark, and Megamouth. These sharks are more concerned with efficiently filtering plankton from as large a volume of water as they sample. Thus, streamlining has been abandoned in favor of a more expansive head and a larger gape.

The final body type of open-ocean sharks belongs to the smaller, more darkly pigmented vertically migrating species, such as the Cookiecutter and Crocodile Sharks. These range from being cigar-shaped to fusiform (tapering at both ends), with a caudal fin that is designed less for speed than for acceleration, and smaller pectoral fins. The prey of these small sharks typically consists of smaller fish and squid, although the Cookiecutter's preference is for larger fish and mammals, from which it extracts a plug of tissue after rapidly accelerating toward its victim. In fact, there are cases of Cookiecutter Sharks biting human swimmers who have attempted to cross deep-sea channels between oceanic islands at night.

Different body types of oceanic sharks

Porbeagles (*Lamna nasus*) are highly-streamlined ocean cruisers with conical heads, deep, robust bodies, and caudal fins built for high speed swimming. The body shape of Oceanic Whitetip Sharks, while still streamlined, is less deep, the head is blunter, and the caudal fin is more asymmetrical, all in line with slower oceanic cruising. The Basking Shark (*Cetorhinus maximus*), with its large head, is less streamlined and has evolved to swim at a slow pace and with a wide gape as it filters plankton. Finally, the tubular-shaped Cookiecutter Shark (*Isistius brasiliensis*), combined with its small pectoral fins and its compact caudal fin, equates to high acceleration more so than speed. Sharks not drawn to scale.

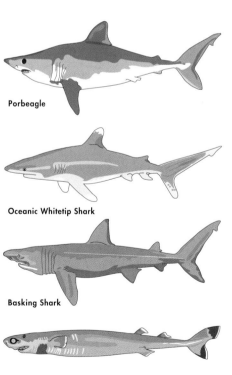

Porbeagle

Oceanic Whitetip Shark

Basking Shark

Cookiecutter Shark

← The Oceanic Whitetip Shark, formerly among the most abundant and widely distributed species of open-ocean shark, has experienced declines in many areas as a result of fishing pressure.

LIFE HISTORIES

Open-ocean sharks also vary from "typical" sharks in some of their life history characteristics, and in turn these vary somewhat among the different body types described above. Although it may not always be clear what the specific advantage of any single life history characteristic might be, it is reasonable to conclude that it might represent adaptations to the specific environmental conditions where the shark lives.

In terms of reproductive output, sharks (and other animals) take one of two paths. In the first, the species produces many offspring in the hope that at least a few will survive, but large litters dictate small birth size, which leads to higher mortality through predation. This is the tactic taken by Blue Sharks and Whale Sharks, which are capable of producing more than 100 and 300 offspring, respectively. In the second path, the species produces very few offspring but these are very large at birth and more likely to survive. Bigeye Threshers usually produce only two offspring, and Crocodile Sharks, Salmon Sharks, and Porbeagles produce just four. Still plenty of other species hedge their bets between the two extreme paths (including Shortfin Mako, Silky, Oceanic Whitetip, and White Sharks), producing litters of 10–20 relatively large offspring.

To summarize and generalize, none of the open-ocean sharks lay eggs (likely because among egg-laying sharks as a group, eggs are always either attached to a substrate or wedged into crevices). Instead, some give birth to large litters of young born at an advanced stage and capable of immediately coping with the exigencies of life at sea. Mortality is high in these young, but they grow quickly to a large size and mature early. In contrast, other oceanic sharks give birth to very few young, but these are larger at birth and hence have a higher chance of escaping predation.

LIFE HISTORY CHARACTERISTICS

Characteristics of oceanic sharks

These sharks, clockwise from top, give birth to live young, are far-ranging, may be sluggish but are capable of high speeds, have numerous young, and are generalist foragers.

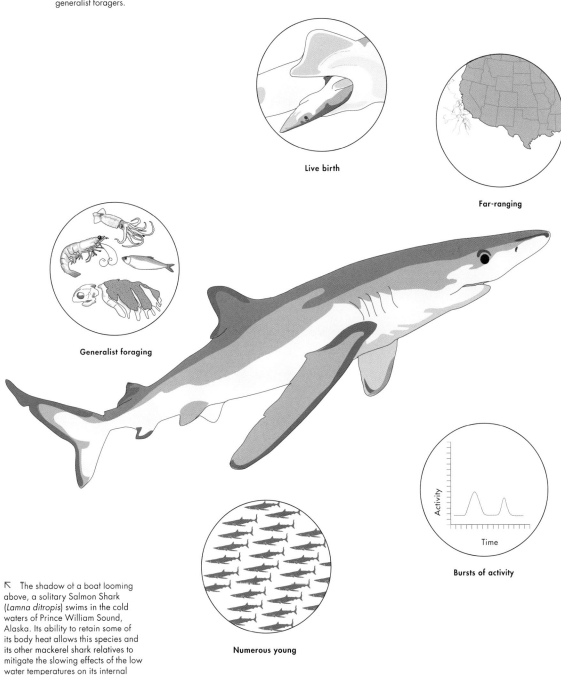

Live birth

Far-ranging

Generalist foraging

Bursts of activity

Numerous young

↖ The shadow of a boat looming above, a solitary Salmon Shark (*Lamna ditropis*) swims in the cold waters of Prince William Sound, Alaska. Its ability to retain some of its body heat allows this species and its other mackerel shark relatives to mitigate the slowing effects of the low water temperatures on its internal function and swimming performance.

131

Threats in the open ocean

Despite inhabiting the open ocean far from land, the sharks found in this environment still face considerable threats, the major one—at least for now—being fishing. In commercial fisheries, sharks are either targeted for fins and/or meat, or are caught as unwanted or unused bycatch and disposed of at sea.

GILLNETTING

Gillnets catch fishes that swim into them and become entangled. Globally, these nets are responsible for more shark (as well as turtle and marine mammal) bycatch mortality than any other gear. High-seas drift gillnet fisheries, such as those targeting flying squid and salmon in the North Pacific, have extremely high rates of shark, marine mammal, and seabird bycatch. Approximately 2 million sharks—primarily Blue and Salmon Sharks—were caught in the squid drift-net fishery in the North Pacific in 1990 alone.

LONGLINING

Pelagic longlines consist of a main line as long as 60 miles (100 km), with more than 1,000 hooks suspended from branch lines. These longlines are not anchored, but are set adrift and marked with high-flyers (floats with a radar reflector and, possibly, a radio transmitter) at the ends. They employ a combination of floats, weighted branch lines and varied branch-line lengths to reach the depths of the targeted species.

Open-ocean longlines targeting pelagic fish such as Mahi-mahi (*Coryphaena hippurus*), Swordfish (*Xiphias gladius*), and tunas have low shark bycatch rates compared to gillnets, but since sharks have conservative life history characteristics, and in many cases reduced potential to rebound from population losses, even a low bycatch rate can be harmful. And the bycatch rate in some fisheries can be high: in the western tropical Pacific tuna fishery, for example, shark bycatch is approximately one shark for every two tunas caught.

← A trawl net bursting with Spiny Dogfish (*Squalus acanthias*), flounder, and cod. In the early 1980s, in response to plummeting U.S. stocks of groundfish like cod off New England, fisheries managers encouraged the development of fisheries targeting Spiny Dogfish to supply "fish and chip" markets in Europe, leading to overfishing of the sharks.

→ A Pelagic Thresher on a commercial longline in the Cocos Islands, Costa Rica. All three species of thresher are prized for their meat, leading to historic overfishing of some stocks.

THREATS IN THE OPEN OCEAN

LONGLINE SHARK BYCATCH SPECIES

Blue Sharks are the dominant species caught in pelagic longline fisheries, followed by Silky Sharks and Oceanic Whitetips, most of which are discarded alive and their survival rate is relatively high. The majority of the sharks that end up in the international fin trade are bycatch in pelagic longline fisheries, and there is concern that these shark species may become depleted. Shortfin Makos and the three species of thresher sharks are of particular concern as they are the most marketable species yet have extremely low reproductive rates. Some pelagic fisheries retain all the sharks taken as bycatch. For example, the Swordfish fishery off Uruguay typically markets more than 95 percent of the large numbers of Blue Sharks caught.

Measures to reduce shark bycatch on pelagic longlines include using monofilament instead of steel leaders and squid instead of fish as bait. Changing the depth of the hooks can also be effective. The majority of pelagic sharks, including Silky (*Carcharhinus falciformis*; page 272) and Oceanic Whitetip Sharks (*Carcharhinus longimanus*; page 140), spend most of their time at depths shallower than 330 ft (100 m). Increasing the depth of hooks to that level or to 500 ft (150 m) has been shown to reduce bycatch rates significantly for these and most other shark species except Blue Sharks (*Prionace glauca*; page 138), Bigeye Threshers (*Alopias superciliosus*), and Shortfin Makos (*Isurus oxyrinchus*; page 74). More research needs to be undertaken to determine the factors influencing shark bycatch rates in pelagic longline fisheries so that additional mitigation measures can be identified.

ANSWERING OUR QUESTION

We now revisit the question posed at the start of this chapter: if you find yourself floating in the middle of the ocean, far from land, what are the chances you will encounter and be eaten by a shark? A number of factors will determine if a shark detects you at sea, including geographic location, time spent in the water, etc. In the 1970s, scuba instructors would tell their students that, if they encountered an Oceanic Whitetip Shark while diving, they should remove their dive knife from its sheath … and sacrifice their dive

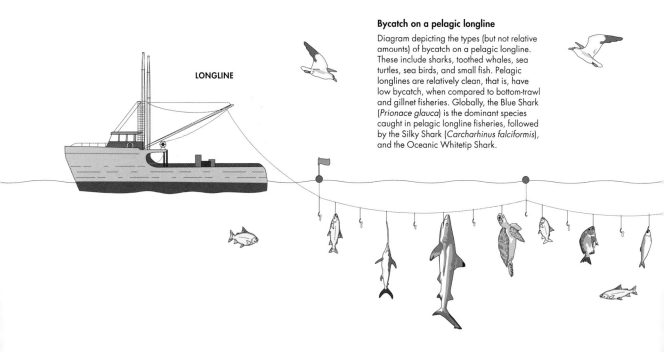

Bycatch on a pelagic longline

Diagram depicting the types (but not relative amounts) of bycatch on a pelagic longline. These include sharks, toothed whales, sea turtles, sea birds, and small fish. Pelagic longlines are relatively clean, that is, have low bycatch, when compared to bottom-trawl and gillnet fisheries. Globally, the Blue Shark (*Prionace glauca*) is the dominant species caught in pelagic longline fisheries, followed by the Silky Shark (*Carcharhinus falciformis*), and the Oceanic Whitetip Shark.

THREATS IN THE OPEN OCEAN

buddy, since there is no escape from the shark. We now know this is hyperbole, and that in fact you would likely have sufficient time to return to your vessel. But there is some truth to the fact that open-ocean sharks are food generalists, and if one does bite you, you can take some small measure of comfort in knowing that it was not a case of mistaken identity. These sharks are naturally inquisitive and will often circle and bump you, apparently to identify your palatability and any threat you might pose. Unfortunately for shipwreck victims, open-ocean sharks are evolutionarily programmed to eat anything they might regard as prey, and that may include humans.

Examples of bycatch in two types of gillnets, drift and set

Globally, gill nets may be responsible for more bycatch mortality of sharks, marine mammals, and sea turtles than any other fishing gear. For example, in 1990 approximately two million sharks, primarily Blue and Salmon Sharks, were caught in the squid drift-net fishery in the North Pacific. Along coastlines, gillnet bycatch includes numerous small species of sharks as well as juvenile Blacktip (*Carcharhinus limbatus*) and Bull Sharks (*C. leucas*), among other species.

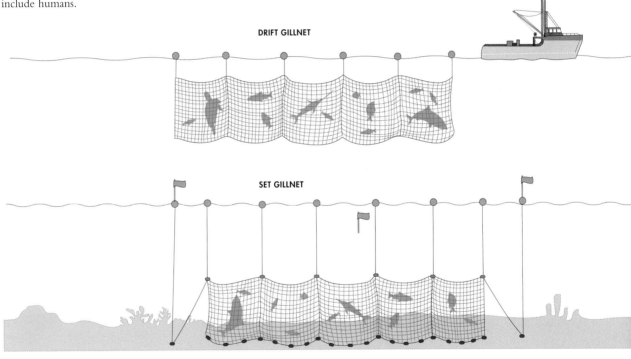

SHARKS OF THE OPEN OCEAN

LAMNA DITROPIS

Salmon Shark

Stout North Pacific salmon lover

SCIENTIFIC NAME	*Lamna ditropis*
FAMILY	Lamnidae
NOTABLE FEATURE	Stout, streamlined shark with a short, conical snout, long gill slits, dusky ventral blotches, and two keels on base of caudal fin
LENGTH	8.5 ft (2.6 m)
TROPHIC LEVEL	Opportunistic top predator of fish—especially salmon, sardines, cod, pollack, and herring—and squid

The Salmon Shark, named for its notable dietary predilection and resulting pinkish flesh, is a sister species and North Pacific ecological counterpart to the Porbeagle. Like the Porbeagle, Salmon Sharks were overfished in the late 20th century. Conservation measures enacted at the cusp of critical population declines reversed the trend, and the species is listed as Least Concern by the IUCN.

Like its close cousins the White Shark, Shortfin Mako, Longfin Mako, and Porbeagle, the Salmon Shark is an oceanic cruiser with more-or-less conical head, deep body, and caudal fin architecture associated with sustained increased power and efficiency at high swimming speeds and efficiency alone at lower ones. The *ditropis* of its scientific name means two keels, and refers to the presence of a strong posterior keel and an additional secondary, reduced keel that extends onto the caudal fin, both of which play roles in swimming performance. In that sense, the Salmon Shark resembles the Porbeagle. However, the Porbeagle has three-cusped teeth (the Salmon Shark has only one additional, small cusplet), and the former has a white blotch at the posterior of its first dorsal fin.

The migrations of the Salmon Shark are noteworthy, in that seasonally they may move between the two most productive ecoregions of the eastern North Pacific, from their summer and fall subarctic habitat to overwintering temperate and subtropical regions (as far south as Hawai'i). Water temperatures in the former are as low as 35°F (2°C) and, in the latter, are as high as 75°F (24°C), representing a very broad range of thermal tolerance. Explanations for their migration include foraging and giving birth.

The North Pacific high seas drift-net fishery for Neon Flying Squid (*Ommastrephes bartramii*), which utilized small mesh nets strung for distances of up to 35 miles (56 km) represented what has been called a "Wall of Death" for Salmon Sharks entrapped as bycatch. A moratorium on drift net fishing enacted in late 1992 was a major step in the recovery of the species.

Although considered an oceanic species, juvenile Salmon Sharks often are found in coastal waters just off beaches, particularly where continental shelves are narrow. Along the California coast, juvenile Salmon Sharks occur so close to shore that dozens become stranded each year. These stranded Salmon Sharks are often confused with juvenile White Sharks due to the conical snout, large black eyes, and starkly defined transition from a nearly black dorsal surface to a white ventral surface.

→ Preparing to devour its namesake prey or simply exercising its jaws, a Salmon Shark moves in its cold Pacific Ocean home.

PRIONACE GLAUCA

Blue Shark

Slender, blue, sluggish, and graceful

SCIENTIFIC NAME	*Prionace glauca*
FAMILY	Carcharhinidae
NOTABLE FEATURE	Relatively slender, countershaded (striking blue on the top and flanks), long snout, teeth not protruding from mouth, long pectoral fins
LENGTH	12.5 ft (3.8 m)
TROPHIC LEVEL	Opportunistic predator of squid and small fish

The Blue Shark is the most abundant and wide-ranging of the open-ocean sharks. It has a placental viviparous mode of reproduction, giving birth to live young that are nourished by an umbilical connection to the mother. Litters are large, ranging from 25 to more than 100, which is uncharacteristic of sharks as a group (Bigeye Threshers, for example, have a litter size of two) but may represent an adaptation to the challenges of an oceanic lifestyle.

Blue Sharks are 1.1–1.5 ft (35–45 cm) long at birth. Like all sharks, there is no parental care, so the young must start life self-sufficient, capable of foraging and avoiding being eaten. As we have already established, this dichotomy is especially worrisome in an environment where potential predators—anything larger than the baby shark itself—are always on the lookout for prey and there simply is no physical refuge in which to hide. In one study of heart function in sharks, a Blue Shark we were operating on gave birth to about 30 near-perfect miniature replicas of adults, many of which emerged snapping (the neonates were all released back into the ocean).

Growth is also atypically fast in Blue Sharks, with males and females reaching maturity at 6–9.3 ft (1.8–2.8 m) in length and 4–5 years of age for males, and 7.2–10.8 ft (2.2–3.3 m) and 5–6 years for females. Male neonates grow an average of about 11–24 in (27–61 cm) per year, which translates into enhanced likelihood of survival.

Blue Sharks are highly mobile and forage on smaller fish and squid, both of whose distributions are generally patchy. Their curved, heavily serrated teeth also allow opportunistic feeding on birds and cetacean carcasses. The sharks are rather sluggish (but still very graceful), and their slender build and large pectoral fins facilitate going with the flow—that is, swimming slowly with oceanic currents, which reduces their energetic costs. While they spend most of their time in the upper 330 ft (100 m) of the open ocean, Blue Sharks are known to dive at night to 3,300 ft (1,000 m) or even deeper. The IUCN lists Blue Sharks as Near Threatened.

→ A Blue Shark off the coast of Rhode Island. Befitting of their high reproductive potential, an anesthetized female Blue Shark on an operating table as part of a study on heart function gave birth to about 30 perfect miniature sharks, all of which swam strongly away after being released back into the ocean.

CARCHARHINUS LONGIMANUS

Oceanic Whitetip Shark

Dominant predator with an undeserved reputation

SCIENTIFIC NAME	*Carcharhinus longimanus*
FAMILY	Carcharhinidae
NOTABLE FEATURE	Stocky; elongate, paddle-shaped pectorals; rounded first dorsal fin; white tips on dorsal, pectoral, and upper caudal fins
LENGTH	Approximately 6.6 ft (2 m) to possibly 11 ft (3.5 m)
TROPHIC LEVEL	Top predator of tunas, Mahi-mahi, other fishes, and squid

The common name of the Oceanic Whitetip Shark refers to the prominent edges of its fins, which although blotchy appear to have been dipped in white paint. In the 1990s, shark scientist Arthur Myrberg conjectured that, from a distance, the white blotches on the fins resembled baitfish, luring fast-moving prey that the shark can then easily capture. Oceanic Whitetips are regularly observed in pairs or small groups, which would likely increase the effectiveness of the fins as bait mimics.

The Oceanic Whitetip Shark is found in temperate and tropical oceans circumglobally. It has a perhaps exaggerated reputation of being an aggressive and dangerous shark, but is also described as sluggish. Given the patchy distribution of prey in the open ocean, all epipelagic sharks necessarily regard any animals (and sometime inanimate objects) as potential food. To assess this possibility, the Oceanic Whitetip Shark is inquisitive and will likely bump people it encounters, and may even bite, so discretion is always advised when diving with this species. Shipwreck survivors report attacks by this species, whose prominent coloration lends credence to these assertions.

Formerly likely as abundant as the Blue Shark or even more so, the Oceanic Whitetip is now categorized as Critically Endangered on the IUCN Red List. Globally, the species is the third most commonly caught shark in pelagic longline fisheries, behind Blue and Silky Sharks. Many of these sharks are discarded live, or finned for the international fin trade and then discarded. Oceanic Whitetip and Silky Sharks are also the dominant shark bycatch species in purse-seine fisheries.

The litters of Oceanic Whitetip Sharks are small, numbering from five to seven. The neonates are 1.8–2.5 ft (55–77 cm) in length at birth.

→ An Oceanic Whitetip Shark, its blotchy fins terminating in its eponymous white tips, orients toward the surface in the clear tropical waters of Cat Island, The Bahamas.

ALOPIAS VULPINUS

Common Thresher

Tail-whipping, weak-jawed leaper

SCIENTIFIC NAME	*Alopias vulpinus*
FAMILY	Alopiidae
NOTABLE FEATURE	Upper lobe of caudal fin nearly as long as body
LENGTH	20 ft (6 m)
TROPHIC LEVEL	Mesopredator of small schooling fishes

The Common Thresher, or simply Thresher, uses its elongated whiplike tail to herd and tighten schools of its prey, which include anchovies, herring, and other small schooling fishes (as well as squid), before swatting and stunning them, and then eating them. These tail-slaps are remarkably coordinated and occur at breakneck speeds of nearly 50 mph (80 kph). Threshers also swat the water, or perhaps swat fish in the water, likely explaining reports by commercial fishers of catching thresher sharks hooked by the tail as they swat the bait.

Threshers live both inshore and in pelagic waters, from the surface down to depths of around 1,150 ft (350 m), in all warm tropical seas. Their diet consists of small pelagic shoaling fish such as anchovies and sardines, and small schooling squid. Consistent with its diet, the mouth of the Thresher is small and its jaws are weak. The closely related Bigeye Thresher (*Alopias superciliosus*) eats slightly larger prey and thus has a somewhat larger mouth and bigger teeth.

The evolution of an elongated upper lobe of the Thresher's caudal fin was accompanied by another anatomical modification: greatly expanded pectoral fins. In fact, the outline of a Thresher from above is vaguely reminiscent of an airplane. Such a modification was required because the asymmetrical tail of sharks in general produces forward and upward thrust at an angle of about 45°. This has the tendency to push the anterior body downward. The snout of the Thresher, unlike its broad, flattened counterpart in many sharks, is conical and does not assist with the required lift. However, the expanded pectorals of the species serve in part to resist this downward force, and the increased size of the fins is roughly in proportion to the elongation of the caudal fin's upper lobe. In this way, the diagonally upward thrust is counterbalanced.

The Common Thresher and Bigeye Thresher are listed as Vulnerable by the IUCN, while the Pelagic Thresher is considered Endangered. The meat and fins of all three species are highly marketable. Threshers are also known for their remarkable leaping ability, in which they completely clear the water, a behavior made possible by their elongated caudal fin and high-performance metabolic machinery.

→ A Common Thresher near Oceanside, California. Once but no longer overfished, Common Threshers of the eastern Pacific, which migrate seasonally between Oregon/Washington to as far south as Baja, Mexico.

SHARKS OF THE OPEN OCEAN

RHINCODON TYPUS

Whale Shark

The largest fish in the sea

SCIENTIFIC NAME	Rhincodon typus
FAMILY	Rhincodontidae
NOTABLE FEATURE	Checkerboard spot pattern, terminal mouth
LENGTH	59 ft (18 m)
TROPHIC LEVEL	Planktivorous, but has also been observed feeding on baitballs of small schooling fishes

In summer, massive numbers of Whale Sharks gather off Mexico's Yucatán Peninsula to feed on dense patches of pelagic fish eggs. This iconic species has also been observed aggregating in about 20 other hotspots in warm waters worldwide, including Australia, the Maldives, Mozambique, and Honduras. These represent extraordinary examples of sharks congregating for food, and such gatherings may also facilitate some exchange of social information that plays a role in mating.

The Whale Shark, the only member in the Rhincodontidae family, is a huge planktivorous, filter-feeding shark with a terminal mouth. It is also the only pelagic species in its order (Orectolobiformes), which includes 42 species such as the Nurse Shark (*Ginglymostoma cirratum*; page 40), bamboo sharks, wobbegongs, and the Zebra Shark (*Stegostoma tigrinum*).

Whale Sharks are widely distributed in tropical and warm-temperate waters, and are one of only three filter-feeding shark species (the others are the Basking Shark and Megamouth). In some parts of the Whale Shark's range, the species is targeted for meat, liver oil, cartilage, and especially its large fins, which are exported.

The species' mode of embryonic nutrition was uncertain until Taiwanese scientists dissected a pregnant Whale Shark that was being sold for meat and found about 300 embryos in egg cases. This mode of reproduction is called aplacental viviparity (formerly termed ovoviviparity) and is practiced by about 40 percent of sharks. Basically, eggs are retained in the mother, hatch internally, and young are born live. Whale Sharks and their orectolobiform cousins the nurse sharks have a thick egg case around the embryo, which distinguishes them from other aplacental viviparous sharks.

Whale Sharks are listed as Endangered globally on the IUCN Red List. There are concerns that tourism at sites where the Whale Sharks congregate to feed and mate could have deleterious effects on numbers by potentially disrupting the species' feeding and social interactions.

→ A Whale Shark in Papua New Guinea. The burgeoning Whale Shark tourism industry raises questions about how to balance economic benefits with safeguarding the species.

SHARKS OF THE DEEP SEA

SHARKS OF THE DEEP SEA

It's cold and dark outside

Not surprisingly, we seldom encounter the small, dark-hued sharks that live their entire lives in the darkness of the deep sea, and, being not particularly scary, they don't garner the media attention of their larger coastal and pelagic counterparts. But in reality, these deep-sea sharks are much more representative of a typical shark than the familiar species of television documentaries.

A HIPPO ON YOUR BIG TOE

At 6,500 ft (2,000 m), hydrostatic pressure equals 2,929 psi (20,194 kPa), which is a bit like having the entire weight of a Hippopotamus standing on your big toe! During our research cruises to study deep-sea sharks, we decorate foam coffee cups commemorating the expedition and send them in cages to the deepest depths we encounter, usually about 6,700 ft (2,000 m), in what is called the bathyal zone. On its return to the surface, a typical foam cup is compressed by the pressure it was subjected to at depth to about a quarter of its original size.

WHAT IS THE DEEP SEA?

Before we dive into an exploration of the sharks that live in the deep sea, let us consider the environment itself more closely. Oceanographers define the deep sea as the areas of the world's oceans that are deeper than 650 ft (200 m). This region is often referred to as the largest ecosystem on the planet, owing to the fact that oceans cover 71 percent of the Earth's surface and the vast majority of this area is deep sea—in fact, 84 percent of the world's ocean habitat is deeper than 6,500 ft (2,000 m). In the zone between 650 ft (200 m) and 3,300 ft (1,000 m), sharks are often the dominant deep-ocean predators, and their abundance and diversity peak at intermediate depths of 1,300–2,600 ft (400–800 m).

→ A bulbous head, small mouth, tapering body, and cartilaginous skeleton are all characteristics of the chimaera, or ghost shark, cousin of sharks and rays. Pictured is the Rabbit Fish (*Chimaera monstrosa*), found in deep water in temperate and polar seas.

SHARKS OF THE DEEP SEA

#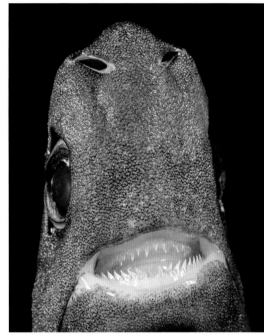

CHALLENGES OF LIFE AT DEPTH

As we descend from the ocean surface to the deep sea, there are three main characteristics that change drastically, particularly over the first 3,300 ft (1,000 m): light levels and temperature decline, and water pressure increases. At depths of 650 ft (200 m) or more, light penetration is insufficient for photosynthesis to take place and light levels are also too low for the human eye to detect, so to us everything is pitch black. However, between 650 ft (200 m) and 3,300 ft (1,000 m), known as the mesopelagic zone or twilight zone, there remains enough light in blue-green wavelengths for many of the inhabitants here to detect and use for vision. Below 3,300 ft (1,000 m) deep, the ocean is almost totally devoid of light.

Temperature also declines with depth in the ocean, although the rate of decline can vary. The temperature of the sea surface depends on where you are in the world, since it varies with air temperature, but in the twilight zone is a permanent thermocline, a transition zone between the warmer upper layer and the cold deep water below. By the time you reach 3,300 ft (1,000 m) in most of the world's oceans, the temperature has generally declined to 36–41°F (2–5°C) and it doesn't get much lower as you go deeper than this.

The third characteristic of oceans that changes with depth is physical pressure. This pressure in the ocean increases by 1 atm with every 33 ft (10 m) of depth, or 14.7 psi (101 kPa).

↑ This might not look like the head of a shark, but the Pygmy Shark (*Euprotomicrus bispinatus*) is as much a shark as is the White Shark (*Carcharodon carcharias*). A vertical migrator that lives in deep water by day and near the surface at night, it reaches a length of about 6 in (15 cm).

↖ A Spotted Ratfish (*Hydrolagus colliei*), an Eastern Pacific chimaera inhabiting benthic environments as deep as 2,970 ft (900 m). Chimaeras are cartilaginous fish that are cousins to sharks and rays.

IT'S COLD AND DARK OUTSIDE

Depth distribution of sharks of the open ocean

Sharks pictured in the epipelagic zone (surface to 660 ft/200 m) include Oceanic Whitetip Sharks, Great Hammerheads, and White Sharks. In the middle, or mesopelagic, zone (660 ft/200 m to 3,300 ft/1,000 m) are found Cookiecutter Sharks, Goblin Sharks, Frilled Sharks, Gulper Sharks, and Cuban Dogfish. In the deep sea below the mesopelagic zone to no deeper than 9,900 ft/3,000 m, part of the bathypelagic zone, Portuguese Dogfish and Bluntnose Sixgill Sharks are shown.

SHARKS OF THE DEEP SEA

Variety among the deep-sea sharks

Two-thirds of all living shark species reach a maximum length of less than 3 ft (1 m), and fewer than 20 percent exceed 5 ft (1.5 m). In addition, about 53 percent of all living shark species and 48 percent of all chondrichthyans (sharks, rays, and chimeras) live their entire lives more than 650 ft (200 m) below the ocean surface. This assemblage includes two primitive sharks: the Frilled Shark (*Chlamydoselachus anguineus*; page 168) and Goblin Shark (*Mitsukurina owstoni*; page 166).

HABITAT DIVERSITY

In the previous chapter we discussed the reasons for the relatively low diversity of sharks in the open ocean, including the homogeneous habitats that limit isolation, and the low productivity and low biomass of prey. You may predict that these same conditions would limit the diversity of deep-sea sharks. After all, there is no productivity through photosynthesis in the deep sea, and the available habitat would seem to be just as expansive and homogeneous as the open ocean. In fact, up until the late 1860s the prevalent theory (dubbed the Azoic hypothesis) predicted that there was no life in the oceans at depths greater than 1,800 ft (550 m). We now know this is far from the case, since more than half of all living shark species inhabit the deep sea.

↗ A rare photo of a living Goblin Shark (*Mitsukurina owstoni*), from Tokyo Bay, Japan. The Goblin Shark lives in deep waters (1,800 ft/550 m) in both the Atlantic and Pacific, where it feeds on fish and squid, and it may undertake daily vertical migrations.

← The head of a Frilled Shark (*Chlamydoselachus anguineus*) with its recurved (pointing backward) rows of very sharp teeth, an adaptation to definitively hold on to any prey item it bites.

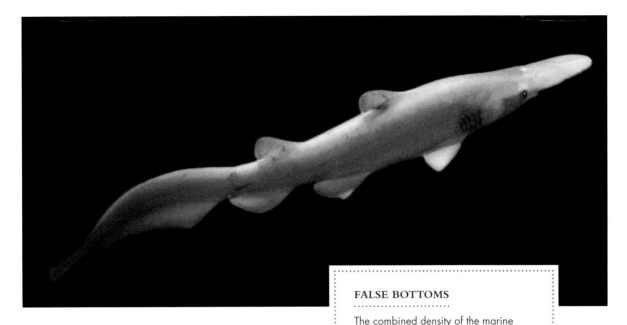

> **FALSE BOTTOMS**
>
> The combined density of the marine animals that migrate vertically each day is so great that when they ascend at night, it appears on a ship's sonar as if the seabed is much shallower than expected. Before it was understood, this phenomenon confused night-sailing ships and submarines.

Since pressure, temperature, salinity, and light levels are similar across and between deep-ocean basins, it is not surprising that many deep-water shark species have very broad, sometimes global, distributions, much like open-ocean sharks. But the deep sea offers far more habitat diversity than the open ocean, which can lead to specialization and isolation, and the evolution of new species. The biomass and diversity of most predators, regardless of ecosystem, tends to be highest on habitat edges, and the deep sea offers some of the most dramatic edge habitats on the planet, including submarine canyons, seamounts, island slopes, and the transition zones between the continental shelf and slope. In our own research into deep-sea sharks, we found that in some submarine canyons the species and communities found on one side are very different from those on the other side, despite being separated by only a short distance.

THE LARGEST ANIMAL MIGRATION ON THE PLANET

As mentioned earlier, photosynthesis in the open ocean occurs in the upper 650 ft (200 m), the so-called epipelagic zone above the thermocline. As water cools at the top of the thermocline, its density increases and causes sinking phytoplankton and zooplankton to become concentrated at an average depth of about 650 ft (200 m). This diel vertical migration, the largest animal migration on the planet, which has been validated as a Guinness world record, occurs on a daily cycle to exploit this concentration of energy in an otherwise oligotrophic (nutrient-deficient) ocean. The migration involves billions of tons of small fishes (especially bioluminescent lanternfish), shrimp, squid, and other taxa, collectively referred to as mesopelagic animals, that ascend from the depths at sunset to feed on the accumulation of zooplankton and phytoplankton. At risk of being seen by predators during daylight, this entire biomass migrates back to depths of around 3,300 ft (1,000 m) at sunrise. They take with them energy in the form of food in their bellies, and may themselves become prey to deep-sea predators, including sharks.

Many deep-sea sharks are bottom-dwelling species that probably rarely venture from the seafloor. This diverse assemblage is dominated by cat sharks (family Scyliorhinidae) deepwater catsharks (Pentanchidae), gulper sharks (Centrophoridae), and dogfishes (Squalidae). Whereas some of these sharks specialize in eating benthic invertebrates such as crabs and worms, many feed at least in part on the fishes, squid, and shrimp that make up the daily mass migration, intercepting them as they dive back to the depths during the day. However, many species of deep-sea sharks participate in this diel vertical migration themselves, either feeding directly on animals that inhabit the mesopelagic zone (as in the case of many lantern sharks, genus *Etmopterus*), or feeding on the predators of the small migrators (as in the case of cookiecutter sharks, in the genus *Isistius*, and sixgill sharks, in the genus *Hexanchus*).

HOW LOW CAN YOU GO?

Many deep-water sharks that were previously considered rare are actually quite common in the ocean depths. However, there appears to be a limit to the depth at which deep-sea sharks can live. Whereas some bony fishes are known to live to more than 26,000 ft (8,000 m) below the surface, the deepest records of a shark are of a Portuguese Dogfish (*Centroscymnus coelolepis*) at about 12,000 ft (3,700 m).

In fact, below about 9,800 ft (3,000 m) the oceans are almost completely devoid of sharks.

There are a few theories to explain the near-complete absence of sharks below 9,800 ft (3,000 m). It may be that food resources are so limited in the deep ocean that larger predators near the top of the food chain cannot find enough prey and would essentially starve or have insufficient energy to reproduce. Another theory is based on shark physiology. All sharks need trimethylamine N-oxide (TMAO) to help maintain their osmotic balance, and they biosynthesize the compound themselves, but in deep-sea species this compound also protects their proteins against damage from the extreme pressure they experience. Below a certain depth, the concentration of TMAO needed to counteract the pressure may exceed a shark's ability to produce the compound, and thus at greater depths osmotic balance cannot be achieved. Many deep-sea bony fish also use TMAO for this purpose, which may explain why so few species occur below 13,000 ft (4,000 m).

↑ The Portuguese Dogfish (*Centroscymnus coelolepis*) is a widely distributed deep-sea shark. It is an active predator of cephalopods and fish, and grows to about 3.3 ft (1 m).

VARIETY AMONG THE DEEP-SEA SHARKS

Diurnal vertical migration

As night transitions into day, an unfathomably large group of mesopelagic small fishes and squid descend into their daytime residence. The previous evening, this assemblage migrated from the depths to the surface to feed, and be fed upon by each other and larger predators like mesopredatory sharks, which follow them. This cycle, the largest animal migration on the planet, plays out daily in the world's oceans.

SHARKS OF THE DEEP SEA

Adaptations of deep-sea sharks

Superficially, many deep-water sharks may seem like small, drab species that are not particularly charismatic, but some of them have specialized adaptations for life in the deep sea that are anything but boring.

→ Contemporary art or a close-up of the skin of a bioluminescing Pygmy Shark? The blue dots are from the photophores (light organs).

↓ Green eyes are indicative of a deep-sea shark capable of absorbing the blue-green light that penetrates to depths of 650–3,280 ft (200–1,000 m). Shown here is the Kitefin, or Seal, Shark (*Dalatias licha*), whose mouth closely resembles that of the Cookiecutter Shark (*Isistius brasiliensis*), with which it very likely shares the same mode of feeding.

LIGHT AND VISION

We have already mentioned sharks that biofluoresce (see Swell Shark, *Cephaloscyllium ventriosum*; page 42), but species in three families of deep-sea sharks actually glow in the dark by producing light, a phenomenon called bioluminescence. The lantern sharks (family Etmopteridae), the kitefin and cookiecutter sharks (Dalatiidae), and some sleeper sharks (Somniosidae) create blue-green light from tiny photophores, small organs that are often arranged in very specific patterns on the ventral surface of the animal. Sharks with photophores use this light to identify other species, conspecifics and, potentially, mates, and it also helps them avoid predators. Because the light they produce is of a similar wavelength to the downwelling light, their silhouette is broken up when viewed by predators from below.

As discussed earlier, visible light attenuates (decreases) rapidly with ocean depth, to the extent that below 3,300 ft (1,000 m) the sunlight filtering down from above is so limited that many fishes, including sharks, have reduced eyes and no longer rely on vision as a major sense. However, in the so-called twilight zone, very dim blue to blue-green light continues to filter through, and the fishes that live at these depths have visual systems adapted to use this light as effectively as possible. Sharks that live in the twilight zone, such as the Icelandic Catshark (*Apristurus laurussonii*), often have large green eyes capable of harvesting this dim light. Whereas this habitat would appear pitch black to the human eye, to the eye of a deep-sea shark it is perfectly lit.

> ### WHAT'S THAT IN YOUR POCKET?
>
> The pocket sharks (genus *Mollisquama*), a group of tiny species that are highly susceptible to being eaten in the deep sea, may take predator avoidance a step further. It is thought that these sharks produce a bioluminescent fluid from glands behind their pectoral fins that is stored in their eponymous pockets. They secrete this luminous cloak as camouflage.

SHARKS OF THE DEEP SEA

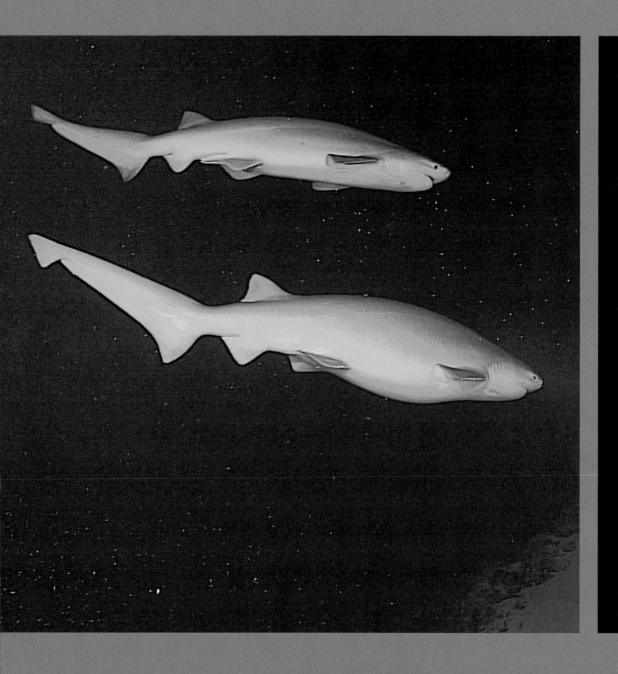

↑ Similar to their ancestors of 50 million years ago, a pair of adult Bluntnose Sixgill Sharks swim together in 1,640 ft (500 m) of water. The photo was taken by a remote camera during the Okeanos Explorer 2017 Pacific Remote Islands Marine National Monument Expedition. The sharks were about 14 ft (4.2 m) long.

↗ The underside of a Pygmy Shark emitting bioluminescent light. Although the shark appears to be illuminated here, the light emitted matches the dwindling downwelling light, so the silhouette of the shark disappears, camouflaging it from potential predators below.

ADAPTATIONS OF DEEP-SEA SHARKS

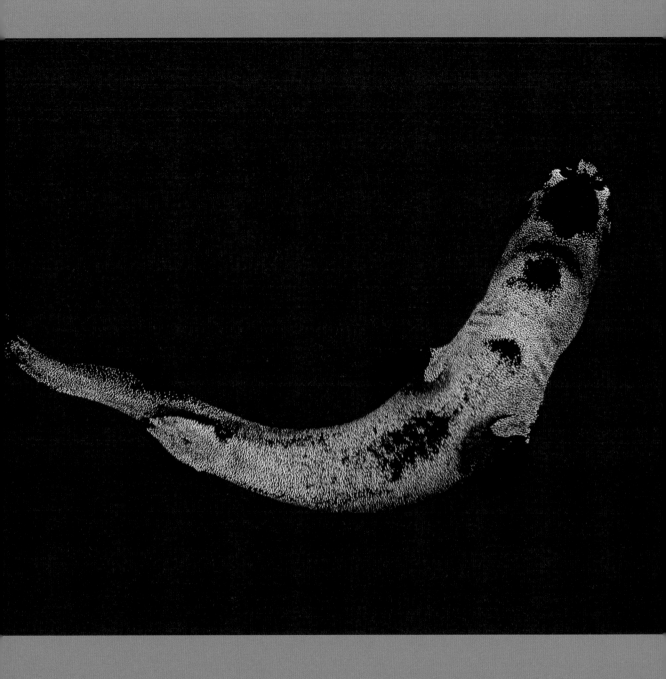

SHARKS OF THE DEEP SEA

ALL THE BETTER TO EAT YOU WITH

In addition to having specialized eyes and photophores, deep-water sharks possess remarkably variable teeth. For example, tiny catsharks have many rows of small yet pointed teeth for grasping invertebrates and fishes and swallowing them whole. The dogfishes and sleeper sharks have teeth with a single functional row that are arranged such that they line up to form a single cutting edge for biting soft-bodied prey such as squid in half, but also allowing them to slice chunks from larger soft-bodied prey. Cow sharks (family Hexanchidae) take this single functional row of teeth to the next level, having very distinctive teeth shaped like a cockscomb. This is an adaptation for sawing pieces of soft tissue out of the carcasses of large animals such as whales, which the sharks may scavenge from the seafloor. Some kitefin sharks have small, narrow-cusped upper teeth and very large triangular lower teeth arranged like a picket fence.

The best known of these, the cookiecutter sharks, use their specialized dentition in concert with fleshy lips to latch onto large marine mammals and fishes. Through suction and with the assistance of the momentum of their prey, they then rotate their bodies to carve out small chunks of flesh, leaving behind crater-type wounds resembling the indentations made by a cookie cutter—hence their common name.

↑ The teeth of the lower jaw of the aptly-named Cookiecutter Shark enable the species to exploit a food resource unused by most deep-sea sharks— larger fish and mammals, from which it removes plugs of flesh.

→ An Atlantic Sixgill Shark (*Hexanchus vitulus*) off Cape Eleuthera, in The Bahamas. The species, like most deep-sea sharks, remains in large part unstudied.

OTHER ADAPTATIONS

Sharks that live in the deep sea have a number of additional changes to their anatomy and physiology, although whether or not these are adaptive is not always clear. Consider the sharks of the genus *Hexanchus*, for example, which have six gills, a single dorsal fin, distinctive teeth, and a long tail. *Hexanchus* is one of the most primitive shark genera, with fossilized examples dating back to the Early Jurassic, 190 million years ago, and has largely occupied the deep sea throughout its evolution. It is thus possible that the evolutionary age and degree of isolation of the genus has resulted in the development of different anatomical and physiological systems.

The constancy of the deep-sea environment also affects its inhabitants. Whereas most coastal sharks exhibit some form of seasonal migratory behavior, many deep-sea sharks that have been examined to date do not. Deep-sea sharks also exhibit continuous rather than seasonal breeding. This involves year-round reproduction, whereby females are pregnant for up to a year or more and vitellogenesis (development of ova) occurs during pregnancy in preparation for mating again. Females typically mate soon after parturition (birth), with no resting phase. Continuous reproduction is characteristic of deep-sea dogfish (order Squaliformes), which are continuously pregnant.

SHARKS OF THE DEEP SEA

Studying deep-sea sharks

Unfortunately, our knowledge of the biology of marine species often lags behind the development of the fisheries that exploit them. This information lag is particularly extreme among deep-sea fisheries that catch sharks.

↑ Deep-sea fishing in the Norwegian Sea on board the trawler *Grande Hermine* in 2011. Sharks are frequent bycatch in deep-sea fisheries.

→ A Pacific Spiny Dogfish (*Squalus suckleyi*) ensnared by a gill net in British Columbia, Canada.

The logistic challenges and expense of studying the deep sea are chief among the constraints of understanding the biology of deep-sea sharks. Aside from species associated with some oceanic islands whose slopes can be reached by small boats, studies of deep-sea sharks often require large vessels with running costs that reach tens of thousands of dollars per day. Even when financial support is available, weather and sea conditions often limit the days available for research. In addition, it is no trivial challenge to develop methods to study deep-sea sharks effectively.

The obstacles to accessing the deep sea have long been overcome by commercial fisheries, which continually move deeper into the world's oceans. As the human population has increased, many coastal fishery resources have become overfished and technological advances in vessels and equipment have enabled fisheries to expand their access to deep-water resources. In some countries, deep-water sharks are targeted specifically for human consumption, but for the most part they are caught unintentionally as bycatch, particularly by large trawl fisheries.

DEEP-SEA DISCOVERIES

Although half of all living sharks are deep-sea species, they are the subject of less than 10 percent of research, in large part because of poor funding. Scientists therefore know little about the basic biology of deep-sea sharks, such as their distribution patterns and life histories, and this limits our ability to manage the fisheries that target deep-sea sharks.

STUDYING DEEP-SEA SHARKS

LIVER LET DIE

Whether targeted specifically or taken as bycatch, deep-sea sharks—particularly the squaliform (dogfish) species—are prized for their liver oil. Often the shark's liver is removed and the rest of the animal is simply discarded.

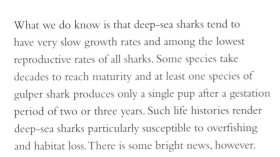

What we do know is that deep-sea sharks tend to have very slow growth rates and among the lowest reproductive rates of all sharks. Some species take decades to reach maturity and at least one species of gulper shark produces only a single pup after a gestation period of two or three years. Such life histories render deep-sea sharks particularly susceptible to overfishing and habitat loss. There is some bright news, however.

There has been a marked increase in the number of studies of deep-sea sharks in recent years, as reflected by the fact that about two-thirds of the new species of sharks discovered in the last 20 years are from the deep sea.

SHARKS OF THE DEEP SEA

ISISTIUS BRASILIENSIS

Cookiecutter Shark

Open-ocean menace

SCIENTIFIC NAME	*Isistius brasiliensis*
FAMILY	Dalatiidae
NOTABLE FEATURE	Cigar-shaped, dorsal fins close to caudal fin, photophores on ventrum except for dark collar near gills, large triangular teeth in lower jaw
LENGTH	1.8 ft (0.5 m)
TROPHIC LEVEL	Opportunistic ambush predator, feeding on chunks of whales, dolphins, and large fishes

The Cookiecutter Shark is a small open-ocean predator found worldwide in temperate and tropical seas that migrates vertically from depths greater than 3,000 ft (900 m) during the day to the sea's surface at night. A member of one of only three families of bioluminescent sharks, this unusual species may use its luminance not only to camouflage itself, but also to lure the larger animals it feeds on.

Being a tiny shark in the open ocean is a dangerous proposition, but Cookiecutter Sharks turn this vulnerability on its head by having among the most specialized modes of feeding of any shark and becoming the predator rather than the prey. When a larger animal such as a Swordfish (*Xiphias gladius*), tuna, dolphin, or whale comes close, the Cookiecutter accelerates toward the unsuspecting victim, then latches on with its sharp triangular teeth and uses its fleshy lips to create suction on the larger animal's flesh. The Cookiecutter is thought to then spin, carving out a chunk of meat or blubber and leaving a cookie-sized crater behind.

Some other members of the family Dalatiidae, including the Kitefin Shark (*Dalatias licha*), have similar teeth and may feed in the same way. For many years it was a mystery what was taking crater-shaped chunks out of the soft rubber on sonar domes of nuclear submarines, but it turns out it was this tiny shark! Cookiecutter Sharks are particularly abundant around the Hawaiian Islands and have even attacked humans swimming in deep water between the islands at night.

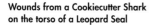

Wounds from a Cookiecutter Shark on the torso of a Leopard Seal
Larger predators such as seals will very likely recover from these wounds, whose diameter is about half the length of a pencil, although the impact of such injuries on their swimming is unknown.

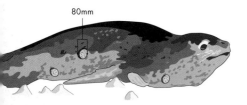

→ A Cookiecutter Shark shown from a view that suggests its alternate common name, Cigar Shark. The bioluminescence on its underside is interrupted by a dark collar without photophores. The darkly pigmented collar could be interpreted from below as a small fish, luring a potential predator to investigate, and possibly becoming the Cookiecutter's prey.

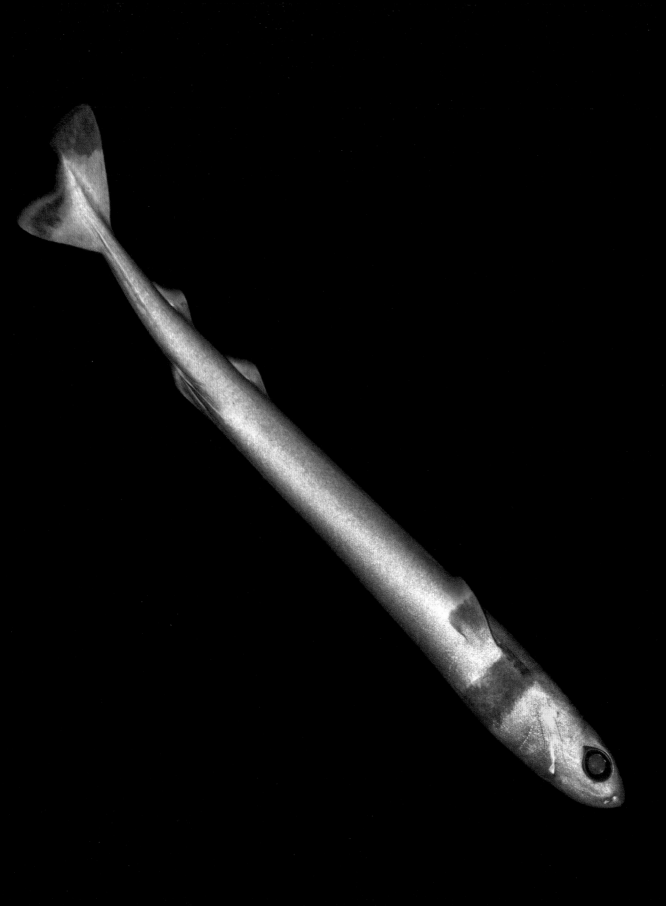

SHARKS OF THE DEEP SEA

MITSUKURINA OWSTONI

Goblin Shark
Paddlefish doppelganger

SCIENTIFIC NAME	*Mitsukurina owstoni*
FAMILY	Mitsukurinidae
NOTABLE FEATURE	Soft-bodied deep dweller, long snout, highly protrusible jaws with dagger-shaped teeth
LENGTH	13 ft (4 m) or more
TROPHIC LEVEL	Deep-sea ambush predator on fishes, squid, and octopuses

Widely distributed but poorly known, Goblin Sharks swim at depths of 900–4,300 ft (270–1,300 m). Slow-moving and nearly neutrally buoyant, they likely rely on stealth to sneak up and ambush their prey. This slow motion and their soft, floppy fins belie the speed at which they extend their highly protrusible jaws to snag their quarry from the water column or off the bottom.

Although most recorded adult Goblin Sharks caught to date have been less than 10 ft (3 m) long, a few estimates from photographs suggest they may get much larger. In most photographs of dead individuals, the sharks are also often seen with their jaws distended, hanging grotesquely. When alive and swimming, the animal is much more hydrodynamic, its jaws retracted into the underslung mouth, which sits beneath a wide blade-like snout unlike that of any of its lamniform relatives. These highly protrusible jaws are a key to the species' survival as a sluggish shark in a deep-sea environment, where prey are few and far between. In what has been called "slingshot feeding," the lower jaw swings downward and back, creating a huge gape and pulling the upper jaw back. This places tension on tendons, which when released leads to the fastest and most extensive forward protrusion measured in any species of shark.

    Little is known about reproduction in the species. Like other lamniform sharks, Goblin Shark embryos are nurtured by eating unfertilized eggs ovulated by the mother.

Extreme eating
The aptly named "slingshot feeding" sees the Goblin Shark's jaws projecting from the mouth at the highest velocity measured in sharks.

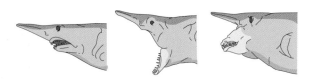

→ A Goblin Shark pictured in the deep sea with a small lanternfish, which it might be stalking.

SHARKS OF THE DEEP SEA

CHLAMYDOSELACHUS ANGUINEUS
Frilled Shark
Blast from the past

SCIENTIFIC NAME	*Chlamydoselachus anguineus*
FAMILY	Chlamydoselachidae
NOTABLE FEATURE	Terminal mouth with uniform rows of recurved trident teeth, one dorsal fin, soft body and fins, six gill slits (the first continuous across the throat)
LENGTH	6.6 ft (2 m)
TROPHIC LEVEL	Deep-sea predator on pelagic squid

The Frilled Shark's species epithet *anguineus* translates to "eel-like." With its long, sinuous eel-like body, terminal mouth (on the end of the snout), amphistylic jaw suspension (upper jaw braced tightly to the skull), and muti-cusped trident teeth, this shark resembles what paleontologists believe some of the ancient sharks of the Devonian and Carboniferous periods may have looked like. Indeed, extinct Frilled Shark relatives date back at least 95 million years and fossils suggest its ancestors date back nearly 400 million years.

The common name of the Frilled Shark refers to the fringed or wavy appearance of the six gill slits. The first of these connects across the throat, creating the appearance of a cloak around the neck—hence the generic name, which is derived from the Greek *chlamydos*, meaning "cloak." Most of what is known of the species comes from occasional catches of one or a few specimens in deep-sea trawl fisheries, although 34 Frilled Sharks were caught in a single trawl on a seamount along the Mid-Atlantic Ridge north of the Azores in 2003. The following year, at a depth of 2,868 ft (874 m), a live Frilled Shark was filmed for the first time in its deep-sea environment during a deep submersible dive on the Blake Plateau off South Carolina.

A study of 264 Frilled Sharks caught in Suruga Bay, Japan, suggested the females are pregnant for three-and-a-half years before giving birth. If true, this is the longest gestation known for any animal.

Frilled Sharks are distributed globally in tropical and temperate waters.

→ With a head more reptilian than shark-like, and a look more ancient than modern, this Frilled Shark specimen is from a depth of over 2,400 ft (730 m) in the Atlantic Ocean.

SHARKS OF THE DEEP SEA

CENTROPHORUS UYATO

Little Gulper Shark

Bring on the slime

SCIENTIFIC NAME	*Centrophorus uyato*
FAMILY	Centrophoridae
NOTABLE FEATURE	Short, pointed spines on the dorsal fins; large greenish eyes; large spiracle; long, free rear tips to pectoral fins
LENGTH	4.2 ft (1.3 m)
TROPHIC LEVEL	Deep-sea mesopredator feeding on squid and fishes such as hakes

The gulper sharks are a family of roughly 16 known small to medium-sized deep-sea shark species distributed around the world. The family has undergone a major revision of its taxonomy in recent years and numerous new species are still awaiting description. The sharks produce copious slimy mucus, perhaps as protection against infection or predation by larger sharks. The Little Gulper Shark is a smaller species, usually found at water depths of 650–3,300 ft (200–1,000 m) in the Atlantic, Pacific, and Indian Oceans, and is the most common species of gulper in the Gulf of Mexico.

Although it is generally considered to be an uncommon species, in specific habitats the Little Gulper Shark can be extremely abundant. In our own research in the eastern Gulf of Mexico, our catch rates of Little Gulpers are higher than for any other deep-sea shark, but they are concentrated in only a relatively small area. At times, on the west side of a single deep submarine canyon, nearly every hook of our survey lines will catch a Little Gulper Shark, yet they are virtually absent throughout the rest of the region. This high density is surprising, considering the Little Gulper Shark has one of the lowest reproductive rates, producing only a single offspring after a gestation period that is believed to be at least two years. This slow reproductive rate, along with the species' tendency to occur in very high densities in specific areas, makes them highly vulnerable to fisheries. The Little Gulper Shark is listed as Endangered on the IUCN Red List.

→ A Little Gulper Shark from southern Australia. Deep-sea sharks, including this species, are adapted to living in a food-poor environment. For example, they have been shown to have smaller hearts than shallow-water counterparts of the same weight.

SHARKS OF THE DEEP SEA

GALEUS MELASTOMUS

Blackmouth Catshark

Abundant egg-laying deepwater cat shark

SCIENTIFIC NAME	*Galeus melastomus*
FAMILY	Pentanchidae
NOTABLE FEATURE	Slender species with jet-black interior of its mouth, marbled pattern of brownish saddles, and prominent saw-toothed, enlarged scales on the upper edge of the caudal fin
LENGTH	1.5 ft (0.5 m)
TROPHIC LEVEL	Generalist mesopredator of crustaceans, cephalopods, and fishes

The Blackmouth Catshark is one of about 110 species in the largest family of all sharks, the deepwater cat sharks. Though diverse, this is also perhaps the least studied of all shark families. Reaching depths of 1.25 miles (2 km), the deepwater cat sharks have adapted to a deep-sea benthic niche that no other sharks and few bony fish exploit. The Blackmouth Catshark occupies a shallower niche than much of its family, between 660–1,600 ft (200–500 m).

The eponymous Blackmouth Catshark, also called the Sawtailed Catshark because it possesses enlarged scales on the anterior portion of the upper lobe of caudal fin, is found in the Northeast Atlantic Ocean and Mediterranean Sea. The Blackmouth Catshark lays eggs like other species in the cat shark families. Egg laying is employed only by small species of sharks (from three orders) that are demersal (living close to the seafloor). Since many eggs can be laid over a long reproductive season, egg laying may be an adaptation that increases the fecundity of small sharks, compared to what would be possible through live birth, given the small body size of the mothers. Blackmouth Catsharks lay eggs year-round, with a peak in winter, and though their fecundity is unknown, it is likely high considering that up to 13 eggs have been observed in the oviduct at one time.

The Blackmouth Catshark possesses adaptations that enhance its ability to locate prey in its deepwater environment, including sweeping its head from side to side while slowly swimming over the bottom, which allows its eyes and ampullae of Lorenzini to sense a wider area of seafloor than if it employed a more typical swimming motion with less head deflection. Its eyes are also adapted to detect light emitted by the bioluminescence of prey organisms, and the enlarged scales on the upper lobe of caudal fin are an adaptation for defense.

The species is listed as Least Concern by the IUCN, even though it is commonly caught as bycatch in deepwater commercial fisheries. The species is of little commercial value.

→ The Blackmouth Catshark was one of the first sharks in which ingested microplastics were documented. Due to the species' generalist, opportunistic feeding habits and relative abundance, researchers are considering using microplastic levels in Blackmouth Catsharks as a way of monitoring microplastics in its environment.

SHARKS OF ESTUARIES & RIVERS

SHARKS OF ESTUARIES AND RIVERS

Estuaries and rivers as shark habitats

Estuaries are highly productive ecosystems and home to more than 50 species of sharks. These sharks may forage, mate, seek refuge, or use the systems as nurseries. At least four shark species, plus members of three ray families, inhabit rivers and lakes. The low salt content of some estuaries and all rivers and lakes presents great physiological challenges as well as ecological opportunities to these elasmobranchs.

Peter Benchley's 1974 novel *Jaws* was apparently inspired by a series of shark attacks that killed four people and injured one along the Jersey Shore in the United States during a 12-day period in the summer of 1916. One attack in the series, in which two people were killed and a third injured, took place in Matawan Creek—on other days a picturesque, peaceful setting— about 1.5 miles (2.5 km) from Raritan Bay and about 15 miles (25 km) from the Atlantic Ocean.

→ Strikingly gorgeous aerial view of an estuary, in Broome, Western Australia, on the Indian Ocean. The region encompasses more than 150 estuaries, potential habitat for a number of sharks, including Bull Sharks (*Carcharhinus leucas*).

176

WHY LIVE IN ESTUARIES, RIVERS, OR LAKES?

During the golden age of sharks 360–300 million years ago (page 34), sharks were the dominant predators not only in the oceans, but also in rivers and lakes. Thus, it should not be surprising that some living sharks inhabit estuaries and rivers. They have not, however, done so continuously since their early evolution, but rather have reinvaded them more recently.

Estuaries are typically highly productive ecosystems that provide sharks with abundant food resources and, for smaller species or juveniles, protection from predation. The advantages of living in freshwater ecosystems are less clear, but could include food resources and protection from predation, as well as decreased competition.

SHARKS OF ESTUARIES AND RIVERS

Historical photographs of Matawan Creek depict a narrow, marsh-lined waterway connected to Raritan Bay. Small creeks such as Matawan are often brackish; that is, they are somewhat diluted by rainfall, runoff, and/or input of fresh water from springs or rivers, and their salinity is lower than that of the open ocean or nearshore environment. Shark attacks that occur far inland where salinities are depressed, such as at Matawan Creek, raise questions about which sharks—especially potentially dangerous sharks— are capable of living in brackish and even freshwater habitats.

We have studied creek systems in the southeast United States similar in size to Matawan Creek, albeit much closer to the ocean, and have captured Bull Sharks (*Carcharhinus leucas*; page 196) and Lemon Sharks (*Negaprion brevirostris*; page 230) similar in size to the White Shark (*Carcharodon carcharias*; page 106) that allegedly perpetrated the 1916 attacks, but we have never seen a White Shark. One reason that has sown doubt that a White Shark was guilty of the attacks is that this species cannot tolerate salinities much lower than that of oceanic seawater, which is high and relatively constant. We will revisit this particular case later in the chapter.

A stylized section through a typical river-dominated estuary

Where fresh water from the river dominates at the extreme left and on the surface, only Bull Sharks are found. At the opposite extreme, on the right, where the salty ocean water is dominant, a mix of coastal shark species is found. In the intermediate area, where lighter river water sits atop denser salt water and may mix with it to variable extents, a mix of shark species occur, but smaller species such as the Atlantic Sharpnose Shark (*Rhizoprionodon terraenovae*) and juveniles such as Sandbar Sharks (*Carcharhinus plumbeus*) may find refuge and abundant food resources there.

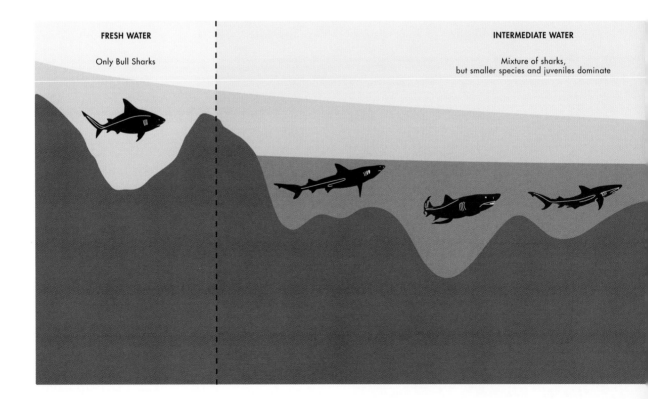

FRESH WATER — Only Bull Sharks

INTERMEDIATE WATER — Mixture of sharks, but smaller species and juveniles dominate

WHAT ABOUT THE RAYS?

The Potamotrygonidae, a family of rays, consists of one marine species and a group of more than 25 mostly landlocked freshwater species living in the Amazon basin. The latter have adapted to their freshwater environment so completely that they are incapable of living in brackish or marine environments. In this group, the rectal glands are vestigial, urea levels are even more reduced, and enzymes and other proteins have lost their sensitivity to urea.

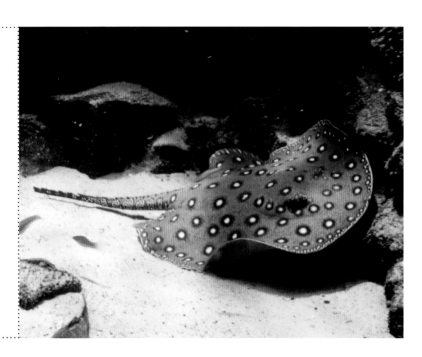

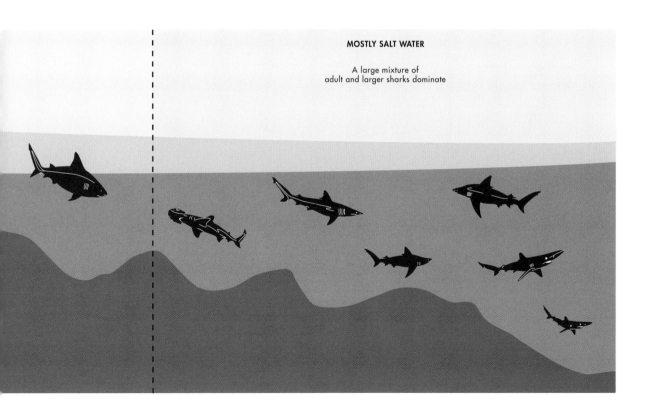

MOSTLY SALT WATER

A large mixture of
adult and larger sharks dominate

HOW CAN SHARKS LIVE IN ESTUARIES AND RIVERS?

Both estuaries and freshwater ecosystems pose physiological problems for sharks, but these are much more extreme in the latter. As a consequence, there are far fewer sharks that are capable of tolerating fresh water.

Let us consider the Bull Shark, the poster child for sharks that can penetrate freshwater habitats. This species is found in almost all of the ecosystems we feature in this book, but it is one of only four known freshwater-tolerant sharks. The other three species are all in the genus *Glyphis*, whose members can be found in the western Indo-Pacific, Papua New Guinea, and northern Australia.

In the nearshore environment (the area over the continental shelf) and in some high-salinity estuaries, Bull Sharks occupy an ecosystem that is saltier than its own interior, which means that the salt (sodium and chloride ions, primarily) in seawater will enter its body through the gills and through the gut when it feeds and drinks. The shark must rid itself of these salt ions or its entire internal physiological system—including its heart, nervous system, and muscles—will cease to function normally. This may prove fatal, since function of the internal environment of sharks—and indeed all vertebrates—relies on a relatively stable salt balance. To maintain the proper equilibrium of salts internally, all sharks, including the Bull Shark, have a small organ called the rectal gland, which expends energy to remove salt from the blood and excrete it into the sea through the animal's cloaca.

↑ After a couple of years, juvenile Lemon Sharks (*Negaprion brevirostris*) are large enough to venture out from the protective cover of the mangroves. Here a 3- or 4-year-old Lemon Shark cruises over the open seagrass and sand bottom of the North Sound of Bimini, The Bahamas.

Dehydration is another potential problem of living in salt water. In sharks as a group, dehydration is avoided by retaining urea, a metabolic waste product. The combination of salt and urea as solutes in the fluids of sharks (for example, blood and tissue fluids) raises the internal solute concentration to approximately equal the external salinity. With the concentration of solutes being about equal inside the shark and in the seawater surrounding it, even if the solutes themselves are not the same, water balance is achieved, and the dehydrating effects of seawater are negated. There is, however, an interesting problem with urea: it is extremely toxic. As a group, sharks have adapted to tolerate the urea in their body fluids, and all of the shark's proteins, including enzymes, not only tolerate the waste product but cannot function in its absence.

For a Bull Shark in a freshwater habitat or a low-salinity estuary, the problems are essentially the reverse. Salts and urea are lost to the environment through the gills and waste products, and water moves in across the gills and gut. These problems cannot be eliminated, but by reducing the salt and urea levels in its body (and by reducing the size of the now unnecessary rectal gland), the Bull Shark and all of the river sharks minimize them.

A BRIDGE TOO FAR

Apparently, the physiological and anatomical adjustments required over time to allow a shark to live in, or even episodically visit, freshwater ecosystems were so extreme that they were evolutionarily a bridge too far for most species. In addition to the problems they create for sharks in terms of salt and water balance, freshwater habitats may impose constraints on the ability of sharks to detect the electrical impulses emitted by their prey using their ampullae of Lorenzini. Other problems include overcoming the reduced viability of sperm in fresh water and living in an environment that might be less stable thermally (since temperature changes more quickly in smaller volumes of water). Sharks may also be somewhat constrained energetically in fresh water, since the absence of salts in the water translates into less buoyancy and thus faster swimming to prevent the shark from sinking. Furthermore, freshwater environments may have limited niche spaces for additional shark species.

Regulating salinity

In salt water, salts diffuse into the bodies of all sharks, including the Bull Shark (pictured), primarily through their gills and when they drink and eat. All sharks must expend energy to pump these salts back into the environment to ensure proper internal function. Reversing these processes in fresh water, where salts are lost from the body and fresh water diffuses in, is possible only in the Bull Sharks and the other three riverine species whose physiological systems have evolved the capacity to both minimize salt loss and water gain, and correct for them.

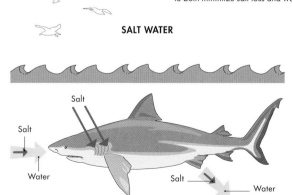

SALT WATER

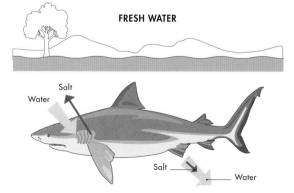

FRESH WATER

SHARKS OF ESTUARIES AND RIVERS

↑ Always looking for a free meal, a resident Nurse Shark (*Ginglymostoma cirratum*) searches for fish scraps around a dock at Cat Island, The Bahamas.

↗ An aerial view of a Bull Shark patrolling the edge of a reef in the Sea of Cortez, Baja California Sur, Mexico.

ESTUARIES AND RIVERS AS SHARK HABITATS

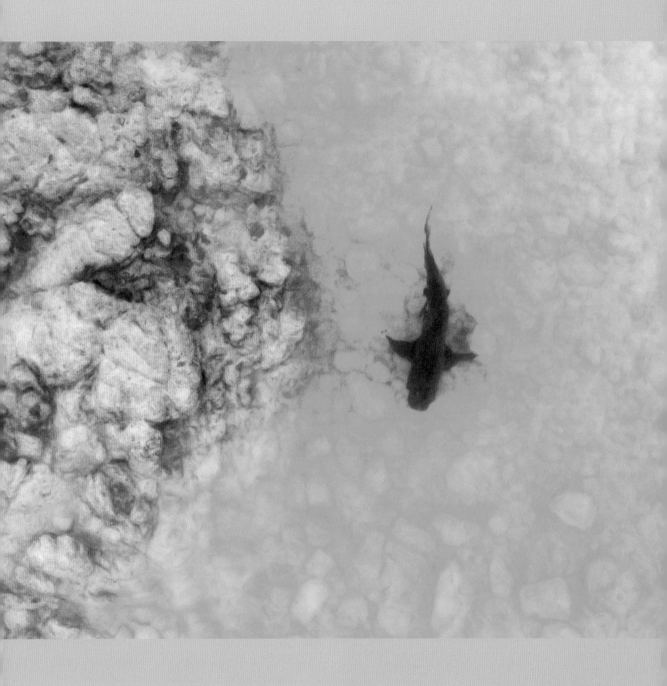

SHARKS OF ESTUARIES AND RIVERS

Who's who among estuarine and riverine sharks

In addition to the four known species of sharks tolerant to fresh water is a much larger number of estuarine species that are euryhaline, or capable of tolerating a wide variety of salinities. This list includes adults of more than 50 species of sharks, as well as juveniles of some species such as Sandbar Sharks (*Carcharhinus plumbeus*; page 198), which may tolerate lower-salinity habitats than adults and in doing so may find refuge from predation from larger sharks, at least temporarily.

Adult sharks found in estuarine systems include many of the more familiar species. They include Bull Sharks, Sandbar Sharks, Lemon Sharks, Atlantic Sharpnose Sharks (*Rhizoprionodon terraenovae*; page 200), Port Jackson Sharks (*Heterodontus portusjacksoni*; page 76), and Bonnetheads (*Sphyrna tiburo*; page 204).

JUVENILE RESIDENTS

In Winyah Bay in the southeast United States, a 25-square-mile (65 km^2) estuary, Sandbar Shark juveniles inhabit lower-salinity areas during low tide (likely for protection from larger sharks and for food) than adults of the same and other shark species. The adults must move either entirely out of estuaries at low tide or into their higher-salinity lower reaches. Chesapeake Bay is likely the largest nursery for Sandbar Sharks in the world, supporting thousands of juveniles. Here, juvenile Sandbar Sharks up to four years old use the lower estuary, those aged 1–3 years use habitats farther in, and the young of the year occur even farther up the estuary in lower-salinity reaches. In terms of their internal physiology, these juveniles behave like Bull Sharks in brackish ecosystems: they lower their internal solute concentrations (primarily sodium, chloride, and urea) to minimize water gain and salt loss. In doing so, their internal fluid environment (blood and other tissue fluids) becomes more like their brackish water habitat, an adaptation that has not been described for most other species of sharks.

VENOMOUS SHARKS IN THE THAMES!

A survey carried out by the Zoological Society of London in 2021 in the River Thames in England recorded three species of sharks, including the Spiny Dogfish (*Squalus acanthias*; page 264 and above). Although the real story was the improved water quality of the Thames, which made the river hospitable again for a number of species that had been absent for a long time, headlines in newspapers and on websites across the globe sensationalized the discovery by focusing on the two venomous spines preceding the dorsal fins of the Spiny Dogfish. While penetration of these spines into your hand can be excruciating, they are only mildly venomous and such injuries are certainly not life-threatening.

↑ A Spiny Dogfish (*Squalus acanthias*), or Spurdog, swims in the cold bottom waters along the coast of Norway. Most members of its families are not coastal, making the Spiny Dogfish the exception. The Pacific Spiny Dogfish (*Squalus suckleyi*) is its ecological counterpart and taxonomic sister species in the north Pacific Ocean.

← Scientists from Coastal Carolina University remove the hook from a juvenile Sandbar Shark caught on an experimental longline in Winyah Bay. The shark will then be measured, tagged, and have a tissue sample removed before being released at the spot where it was captured.

SHARKS OF ESTUARIES AND RIVERS

Ecology of sharks in estuaries and rivers

Because of the proximity of estuaries to human habitats, estuarine sharks are better studied than are sharks in most other ecosystems. The same is not true for sharks in rivers and lakes, in part because there are fewer species in these systems and also because many of the rivers and lakes that support sharks are often in remote areas that are hard to access.

↑ Canada's St. Lawrence Estuary, one of the world's largest and deepest estuaries, is home to at least eight species of sharks, some of which do not typically occupy such systems, including the Greenland (*Somniosus microcephalus*), Blue (*Prionace glauca*), Basking (*Cetorhinus maximus*), and White Sharks (*Carcharodon carcharias*), and Porbeagle (*Lamna nasus*).

← A surprising estuarine inhabitant, this large Greenland Shark cruises along the anemone-covered bottom of the St. Lawrence River estuary along the Atlantic Canadian coast.

Estuaries vary in how they were formed, their age, their physical and chemical characteristics (for example, their size and salinity regime), and their degree of urbanization and other human impacts. Sharks inhabiting these systems may do so year-round, as in the case of the Greenland Sharks (*Somniosus microcephalus*; page 202) in Canada's Gulf of St. Lawrence, the world's largest estuary. Alternatively, they may occupy an estuarine ecosystem on shorter timescales, including seasonally or tidally. In Chesapeake Bay, the largest estuary in the United States, Sandbar Sharks under three years of age migrate south to North Carolina in winter, then move back to the bay when the water warms in the summer.

SHARKS OF ESTUARIES AND RIVERS

Bull Shark

Juvenile Bull Shark

Sandbar Shark

Juvenile Sandbar Shark

Atlantic Sharpnose Shark

Juvenile Atlantic Sharpnose Shark

Blacktip Shark

Finetooth Shark

Blacknose Shark

Lemon Shark

Sections of the two sections of Winyah Bay, South Carolina.

In the middle bay (top), large numbers of juvenile Sandbar Sharks are present. The low-to-intermediate salinities of the middle bay provide ample foraging opportunities for the juvenile sharks as well as refuge from most predators. The big exception is the Bull Shark, which, along with these juvenile Sandbar Sharks, can tolerate these conditions and can prey on some of the Sandbar Sharks. In the lower bay (bottom), a mix of seven or more species of sharks inhabits the system, which is characterized by high, stable salinities, which most of these sharks require full-time. Salinity thus plays a major role in habitat selection in estuarine sharks.

→ A juvenile Sandbar Shark is measured and tagged prior to being released in Winyah Bay, South Carolina.

ESTUARIES AS SHARK NURSERIES

Since the Sandbar Shark is found circumglobally in temperate coastal waters, including estuaries, let us continue to focus on this species, which is also called the Brown Shark. Along the East Coast of the United States, adult Sandbar Sharks mostly reside on the continental shelf at depths of 66–130 ft (20–40 m), but they may move into shallower coastal and estuarine waters seasonally to forage.

Estuaries provide critical nursery habitat for growth and development of juvenile Sandbar Sharks, providing them with abundant food resources (for example, small fish, stingrays, neonate Atlantic Sharpnose Sharks, and crustaceans) and protection from predation, predominantly by larger sharks. Although the largest summer nursery for Sandbar Sharks is found in Chesapeake Bay, there are other nurseries on the East Coast of the United States from New Jersey to at least South Carolina, as well as similar environments elsewhere in the world, such as Turkey's Gulf of Gökova in the eastern Mediterranean Sea. Neonate and juvenile Sandbar Sharks remain in or near the estuaries in which they were born until shortening day lengths signal coming cooler water temperatures and initiate migrations to overwintering areas.

Juvenile Sandbar Sharks in Chesapeake Bay are essentially nomadic, covering on average 15–77 square miles (40–200 km^2) each day, with no defined home range. This adaptation allows them to target the abundant and highly mobile distribution of prey in the bay, such as sciaenid fishes (drums and croakers) and Blue Crabs (*Callinectes sapidus*). In Winyah Bay, a much smaller system, the juveniles tend to move with the incoming tide, potentially keeping a step ahead of their predators—the larger sharks entering the bay—by spreading out and entering the higher reaches. In these parts of the bay, some of the larger sharks are incapable of tolerating the physiological stress imposed by the decreased salinity.

WHERE TO LIVE?

How sharks select and use habitats within estuaries remains a major area of research. Let us again consider Winyah Bay. In the warmer months, eight species of sharks in an array of sizes and ages are commonly found in the bay. Larger species (greater than 5.7 ft/1.75 m) include Lemon Sharks (greater than 10 ft/3 m), Bull Sharks, Sandbar Sharks, and Blacktip Sharks (*Carcharhinus limbatus*). Adults of smaller species, including Blacknose Sharks (*C. acronotus*), Finetooth Sharks (*C. isodon*), Bonnetheads, and Spinner Sharks (*C. brevipinna*), are seasonally present. Among neonate and juvenile life stages in Winyah Bay, Sandbar and Atlantic Sharpnose Sharks (as small as 11 in/29 cm) dominate, but immature individuals of all of these species have been caught here. These species all leave the estuary as day length shortens in late fall, and are replaced in winter by Spiny Dogfish and, occasionally, Dusky Smoothhounds (*Mustelus canis*; page 266).

All of this raises the question of how such a large number and variety of apex predators and mesopredators coexist in such a small ecosystem. Is Winyah Bay simply an alphabet soup, with neonates, juveniles, and adults randomly distributed, and with life or death dependent on whom you encounter and whether they be friend or foe?

The picture that has emerged after two decades of studying Winyah Bay is quite complex (and thus our explanation here is an oversimplification). Three major abiotic factors account for where any shark might be at any time in the bay: tide, salinity, and depth. Other factors—such as time of day, currents, dissolved oxygen levels, and turbidity (degree of cloudiness)—also affect the distribution of sharks, as do mating or birthing if they occur in the bay. Finally, the status and health of the entire biological community of the bay (for example, the abundance and distribution of prey and predators) also plays a major role.

Ask any fisher and they will tell you that larger sharks and bony fishes enter and penetrate estuaries during the flood tide. They then forage in the estuary before riding the currents of the ebbing tide back to its mouth or into the ocean. The sharks (except, of course, Bull Sharks) may be physiologically intolerant of the low salinities at low tide, and they may also prefer to avoid shallower depths. Even at high tides, the larger species seem to favor the deeper sections of Winyah Bay. Many of the smaller shark species—especially neonates and juvenile stages—are potential prey for the larger sharks when these are foraging at high tide. As you might have guessed, these smaller individuals typically move into the more protected shallow waters at the saltmarsh-lined fringes of the system, or move as far upstream as their tolerance of fresh water will allow.

In Apalachicola Bay in northwest Florida, where Bull Sharks are seasonally resident, small sharks must balance the risk of predation versus access to prey. In this estuary, Bonnetheads err on the side of finding prey at night and avoiding predators during day. Yes, some will inevitably succumb to being prey to the Bull, Sandbar, Lemon, and other sharks, but that is part of the evolutionary rhythm of life and the continuing saga of the predator–prey arms race.

BULL SHARK IMPERSONATORS

Tide and salinity are inexorably linked in estuaries such as Winyah and Apalachicola Bays, where river flow of fresh water from the upper reaches of the system battles tidal influx of salt water from the lower reaches to determine the salinity regime. Neonate and juvenile Sandbar Sharks have figured out, evolutionarily speaking, how to live in the slightly fresher waters that are physiologically off limits to the larger sharks—except their close relatives the Bull Sharks, that would gladly eat them if they could. The juvenile Sandbar Sharks mimic the way the Bull Sharks adjust to less salty waters by lowering their own internal solute concentrations.

ECOLOGY OF SHARKS IN ESTUARIES AND RIVERS

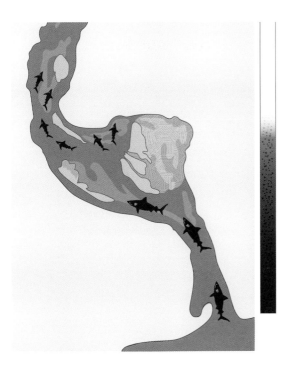

Maps of Winyah Bay showing how tide influences the distribution of sharks

At the top of each map are the rivers that flow into the bay. At the bottom is the inlet opening into the ocean. The bar at right shows salinity in the bay; dark equals high values. At low tide (lower image), most of the adult sharks, as many as eight or nine species, remain in the salty lower bay or nearshore waters since they cannot tolerate the lower salinity of the middle bay. Juvenile Sandbar Sharks use the middle bay as nursery grounds at low tide, where they are protected from their larger shark predators. At high tide (upper image), the incoming tide elevates the salinity in the middle bay, allowing the adult sharks to penetrate further up the bay. In response, the juvenile Sandbar Sharks move to the fringes of the middle bay, or even further up the bay where the salinities remain too low for their adult shark predators. Winyah Bay is an important nursery ground for Sandbar Sharks. The sharks are not drawn to scale.

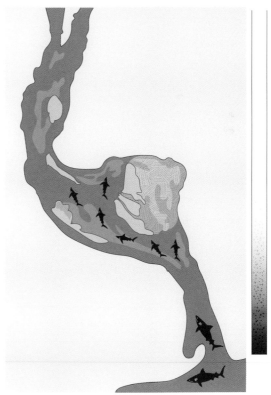

(Sharks not drawn to scale)

SHARKS OF ESTUARIES AND RIVERS

Threatened on all fronts

Not surprisingly, human development along estuaries and freshwater systems, as well as in their watersheds, impacts the sharks in these systems. At least 60 percent of the planet's human population now lives within 62 miles (100 km) of the coastline, and that figure is continuing to grow rapidly. As a result, more than half of the world's coastlines are at significant risk from associated development. Coastal cities, many of which are located on rivers and estuaries, are also reaching sizes unprecedented in human history.

The list of human impacts on estuarine and freshwater ecosystems is long, and includes pollution from agricultural waste and undertreated human sewage, which can cause hypoxia (depleted dissolved oxygen); increased salinity due to removal of fresh water for human uses; micro- and macroplastic and other types of pollution, including other chemicals, noise, radioactivity, and thermal pollution; habitat alteration and destruction; sediment runoff; loss of biodiversity in ecosystems; overfishing; and, of course, climate change, which affects water temperature, precipitation (and thus salinity, as well as water flow rates), biodiversity, acidity, and sea level.

← Industrialization, in this case an oil refinery, surrounds Liverpool's Mersey Estuary, long considered one of the most polluted in the region until efforts to restore it began about 30 years ago. A small number of sharks may inhabit the lower estuary, including Small-spotted Catsharks (*Scyliorhinus canicula*) and Spiny Dogfish.

THREATENED ON ALL FRONTS

Human impacts on an estuary
Although the activities depicted above might not be applicable to every estuary, their presence in most estuary ecosystems represents a veritable minefield for sharks and other organisms.

SO NEAR AND YET SO FAR

Murrells Inlet, an estuary in northeast South Carolina heavily impacted by human activities, has a depauperate shark fauna, with few caught during three censuses in the last decade. North Inlet, 20 miles (32 km) to the south, is a relatively pristine environment teeming with at least eight shark species. Explanations for the disparity remain only hypotheses, including habitat alteration, a disrupted food web, and pollution from noise, nutrients, and chemicals (perhaps polycyclic aromatic hydrocarbons, which are found in Murrells Inlet). Older recreational fishers report that they frequently caught sharks in Murrrells Inlet in the 1970s.

FISHING THREATS

Overfishing, both as bycatch and targeted species, still represents the most immediate concern for several species of estuarine shark, even though most of the fishing is concentrated outside estuarine ecosystems. Of particular concern are the river sharks of the genus *Glyphis*, a group of three poorly known species, all of which are considered threatened or endangered. Small-scale artisanal fishers in fishing villages along estuaries and tidal rivers in the Indo-West Pacific, including Papua New Guinea and northern Australia, catch river sharks in their gillnets, which are difficult to regulate. These small-scale fisheries can have outsized conservation implications for species that are endemic or have small regional distributions, such as the river sharks and freshwater rays, especially when key aspects of their biology and life histories have not been studied.

CLIMATE CHANGE

Sea-level changes represent another growing threat to coastal ecosystems. In a historical context, the sea level has risen about 400 ft (120 m) in the past 18,000 years, and many coastal shorelines have moved far inland. Off the East Coast of the United States, for example, the shoreline has migrated inland more than 50 miles (80 km) during this period. As the rate of sea-level rise accelerates, intertidal and some shallow subtidal marine communities may not be able to keep pace and could

THREATENED ON ALL FRONTS

out how climate change impacts leading to alterations in the shark's habitat would affect the foraging ecology of this life stage. Older juveniles, aged 4–6.5 years, were found to prey primarily on fish in the lower saltmarsh-dominated part of the estuary. As these saltmarsh habitats become inundated by sea-level rise, juvenile Bull Sharks will likely be forced to migrate into areas lacking the protection provided to them by the saltmarsh habitat. Increased temperatures associated with climate change may also ultimately affect the metabolism, swimming performance, and even migratory destinations and species distributions of Bull Sharks and other marine species. Along the east coast of Australia and in the southeast United States, a shift poleward in the distribution of Bull Sharks has been documented and, in the former, the sharks landed in more urbanized, and thus more polluted, estuaries.

THE MATAWAN CREEK CULPRIT

We now revisit the case of the 1916 shark attacks in Matawan Creek, where a White Shark (*Carcharodon carcharias*; page 106) was originally blamed. Based on the evidence, however, it is at least equally plausible that a Bull Shark penetrated into the small estuarine system.

If, in fact, Matawan Creek was brackish (diluted in part by fresh water) at the time, like estuaries may be when rainfall has been high and tidal amplitude has been low, then the big shark in the creek described by witnesses and the species whose biology would allow it to enter what was presumably a less salty system is likely to have been a Bull Shark, a species documented to bite humans. Large, mature White Sharks—that is, those greater than about 15 ft (4.6 m) in length—have never been found in confined systems that far from the ocean. Moreover, adults of this species are very active and require significant space to move. Small creek systems are also not typical habitat for juvenile White Sharks. However, given enough time even extremely unlikely events occur, and firsthand descriptions of the shark in Matawan Creek are consistent with it being a juvenile White Shark, although we will never know for certain.

die off, with repercussions throughout the food web. Such change is now occurring in intertidal saltmarsh environments around the Mississippi River delta in Louisiana, as coastal subsidence—combined with global sea-level rise—exceeds the rate of vertical growth of the marsh community on which the entire ecosystem depends. In other estuaries, mangrove, saltmarsh, and seagrass habitats are being lost as sea level rises, temperatures increase, and more frequent storms dilute the water (that is, lower the salinity).

A recent study on juvenile Bull Sharks (*Carcharhinus leucas*; page 196)—a species dependent on estuaries and rivers in its early life stages—in the Clarence River in New South Wales, Australia, pointed

SHARKS OF ESTUARIES AND RIVERS

CARCHARHINUS LEUCAS
Bull Shark
Lover of fresh water

SCIENTIFIC NAME	*Carcharhinus leucas*
FAMILY	Carcharhinidae
NOTABLE FEATURE	Stout, broad head with short snout, small eyes; serrated triangular upper teeth; no ridge between first and second dorsal fins
LENGTH	11 ft (3.4 m)
TROPHIC LEVEL	Top predator of sharks and bony fishes

The Bull Shark is an impressive and imposing beast, its massive size, solid build, and broad, blunt snout giving it a commanding presence. The species is considered dangerous, especially in its brackish and freshwater habitats in the developing world, where humans and Bull Sharks often overlap in the same water.

In addition to penetrating freshwater habitats, Bull Sharks live in shallow tropical and temperate waters worldwide, in depths less than 100 ft (30 m) to as deep as 500 ft (150 m). Their upper teeth are broad and serrated, adaptations for shearing flesh from their prey.

The Bull Shark is one of the three most dangerous sharks (the others being the White Shark and Tiger Shark, page 232) and quite possibly tops the list in terms of fatal attacks on humans. Local residents and subsistence and artisanal fishers in developing countries coexist peacefully with Bull Sharks in the latter's often murky brackish and freshwater habitats, but occasional attacks do occur, and the number of these is likely underreported. We catch Bull Sharks in our shark surveys in coastal and estuarine waters near crowded beaches along the southern United States coastline, and attacks occur there as well, but they are infrequent. In clear, shallow tropical waters, a Bull Shark that encounters a boat, or perhaps a diver moving toward it, may go into an agonistic display to warn off the perceived threat, lowering its fins and hunching its back. When that occurs, the prudent action is to withdraw, or the shark may charge and strike you or the boat.

Bull Sharks are classed as Vulnerable on the IUCN Red List, and are impacted by habitat degradation, pollution (including climate change), and overfishing. Interestingly, they appear to thrive in areas altered by dredging, marinas, and outfalls from power plants. In one study, Bull Sharks in polluted waters in Florida were found to contain concentrations of six antidepressant drugs (of eight tested). The levels of these antidepressants were considered low and not a health risk to the sharks in the short term, but their presence is a serious concern nonetheless.

→ Multiple Bull Sharks compete for fish scraps in The Bahamas. Large, girthy bodies and strong jaws filled with serrated triangular teeth allow Bull Sharks to be the top predators in many tropical coastal environments.

CARCHARHINUS PLUMBEUS

Sandbar Shark

Graceful gray migrators

SCIENTIFIC NAME	*Carcharhinus plumbeus*
FAMILY	Carcharhinidae
NOTABLE FEATURE	Stout shark, oversized first dorsal fin far forward on body, prominent interdorsal ridge
LENGTH	8 ft (2.4 m)
TROPHIC LEVEL	Mesopredator of small bony fishes and occasionally sharks, rays, crustaceans, and cephalopods (squid and octopuses)

We are often asked what our favorite shark is, a question akin to naming your favorite child. One shark that is always in contention is the Sandbar Shark. With its enlarged first dorsal fin, somewhat docile nature, swimming grace, and amenability to being handled by shark biologists, how could we not admire this species?

The Sandbar Shark is a bottom-dwelling coastal species found worldwide in temperate waters shallower than 330 ft (100 m). It is the dominant shark along the East Coast of the United States and in Hawai'i, but it occupies deeper waters in the latter. It has a small ridge on its dorsal surface between the first and second dorsal fins, and is thus considered a ridgeback shark by fishery biologists. Other common ridgeback sharks in the United States and Caribbean include the Dusky Shark (*Carcharhinus obscurus*; page 270), Silky Shark (*C. falciformis*; page 272), Bignose Shark (*C. altimus*), Night Shark (*C. signatus*), and Caribbean Reef Shark (*C. perezi*; page 234), all of which are prohibited in the United States from being kept by commercial and recreational fishers. Catches of Tiger and Oceanic Whitetip (*C. longimanus*; page 140) Sharks, and Dusky Smoothhounds, also ridgebacks, are limited but not prohibited.

Sandbar Sharks take about 15 years to reach maturity. They produce about eight pups every two years, a conservative reproductive strategy that hinders their ability to recover quickly from historical overfishing. The species has been prized by the shark fin soup trade because of its enlarged first dorsal fin, and as a result of overfishing it is now categorized as Endangered on the IUCN Red List. Despite aggressive management of the targeted longline fishery for sharks along the United States East Coast, both Sandbar Sharks and Dusky Sharks have failed to regain their numbers fully, although the trajectory of their recovery is trending in the right direction.

→ The very large first dorsal and pectoral fins are apparent as this Sandbar Shark glides through the deep blue waters of Hawai'i in the central Pacific Ocean.

SHARKS OF ESTUARIES AND RIVERS

RHIZOPRIONODON TERRAENOVAE

Atlantic Sharpnose Shark

Ubiquitous predator of small fishes and invertebrates

SCIENTIFIC NAME	*Rhizoprionodon terraenovae*
FAMILY	Carcharhinidae
NOTABLE FEATURE	Small black second dorsal fin, small ridge in front of anal fin (pre-anal ridge), elongated ampullary pores resembling oval slits behind eyes on cheeks
LENGTH	3.3 ft (1 m)
TROPHIC LEVEL	Mesopredator of small fish, crustaceans, worms, and mollusks

Atlantic Sharpnose Sharks are hugely abundant in their coastal northwestern Atlantic habitat, and may be beneficiaries of declines in the populations of their main predators, larger sharks. From April or May and throughout the summer, neonate Atlantic Sharpnose Sharks are nearly ubiquitous in temperate estuaries and nearshore environments within their natural range.

There are two stocks of Atlantic Sharpnose Shark, in the northwest Atlantic and the Gulf of Mexico, and neither is overfished. Targeted commercial landings in these two locations, for meat and bait for larger sharks, were 294,000 lb (133,000 kg) in 2020, a value of US$244,000. Recreational fishers caught an additional 196,000 lb (89,000 kg) in the same year.

The fast life history characteristics of the Atlantic Sharpnose Shark, unlike those of Sandbar Sharks, help protect it from the threats of overfishing, even though it experiences extremely high bycatch mortality in United States trawl fisheries. Atlantic Sharpnose Sharks mature at about three years. The litter size is only about four, after a 10–11-month gestation period. Neonates resemble miniature adults. They are born as small as 11 in (28 cm) and grow about 2 in (5 cm) per month over the next three months.

In 2018, we evaluated microplastics in 16 adult Atlantic Sharpnose Sharks from the northeast coast of South Carolina, and found 34–75 particles per individual in 100 percent of sharks examined. This is among the highest reported values for microplastic ingestion among sharks. Another study of microplastics in neonates in the same area in 2020 enumerated 16–63 particles per individual in all 30 sharks examined.

→ An Atlantic Sharpnose Shark off the coast of Mississippi in the Gulf of Mexico. Notice the large nasal papillae (external flap of the nares) and long labial furrows (folds in the corner of the mouth) that are diagnostic of this species.

SHARKS OF ESTUARIES AND RIVERS

SOMNIOSUS MICROCEPHALUS

Greenland Shark

A very old vertebrate

SCIENTIFIC NAME	*Somniosus microcephalus*
FAMILY	Somniosidae
NOTABLE FEATURE	Massive size, cylindrical body, two small dorsal fins, small deep-green eyes
LENGTH	23 ft (7 m)
TROPHIC LEVEL	Top predator of mammals, fishes, and invertebrates

Although the Greenland Shark is considered a deep-water and oceanic species, it is also found in the St. Lawrence Estuary at depths of less than 100 ft (30 m), which is why it is included in this chapter. It is the second-largest carnivorous shark and the largest high-latitude shark.

Videos of the Greenland Shark depict a sluggish, gentle giant, but the species can also be the opposite of this. Its diet consists of active prey, and Polar Bears (*Ursus maritimus*), Moose (*Alces alces*), horses, and Reindeer (*Rangifer tarandus*) have been found in its digestive tract. Individuals have been reported to "stalk" scuba divers, although the sharks may have been innocently following the divers.

Two scientific discoveries in the last decade merit mentioning. First, in 2013 a specimen was caught at a depth of about 5,700 ft (1,750 m) in the Gulf of Mexico. Although this was one of the first observations of the species outside the Arctic, it was not that surprising since the water temperature at that depth in the Gulf of Mexico—about 39°F (4°C)—was close to that in its polar home.

The second discovery was somewhat more remarkable. In 2016, radiocarbon dating was used to estimate a mean age of 392 (± 120) years for one specimen, endowing the species with the age record for a vertebrate. (The oldest living animal is a quahog clam, *Mercenaria mercenaria*, at about 507 years.) Sexual maturity in Greenland Sharks is estimated to be about 150 years, and they have a growth rate of about ⅖ in (1 cm) per year. Living to such a ripe old age comes with a cost, however: potential continuous long-term exposure to environmental toxicants. One study found DDT levels in 15 Greenland Sharks caught in 2010 to be abnormally high. This could indicate that the sharks were exposed to the pesticide between the 1940s and 1970s, when it was most widely used, and that the lower metabolism of the species translated into slow depuration (declines in the chemical due to the shark's metabolism or other means).

→ The Greenland Shark occurs farther north into Arctic Seas than any other shark species, and it also likely has among the widest depth distributions, occurring from shallow boreal estuaries to ocean depths of over 7,000 ft (2,100 m).

SPHYRNA TIBURO

Bonnethead
Speckled crab specialist

SCIENTIFIC NAME	*Sphyrna tiburo*
FAMILY	Sphyrnidae
NOTABLE FEATURE	Small shovel-shaped head, spotted dorsal surface
LENGTH	5 ft (1.5 m)
TROPHIC LEVEL	Mesopredator of small fishes and invertebrates, especially crabs and shrimp

The Bonnethead is an extremely common shark in the Gulf of Mexico and the Atlantic coastline of the United States. It is the only shark whose head shape differs between the sexes: the cephalofoil of females is rounded, whereas males exhibit an anterior bulge.

Shrimp trawl fisheries in the United States catch hundreds of thousands of Bonnetheads and other small coastal sharks, including Atlantic Sharpnose and Blacknose Sharks, to the extent that this bycatch exceeds other sources of fishing mortality for these species. However, the Bonnethead's life history characteristics help to protect it from potential overfishing: it takes three years to mature, reproduces every year, and produces eight or more offspring. Moreover, the Bonnethead is of low economic value, although it is sold for its meat in some markets.

Studies on Bonnetheads in 2015 and 2018 reported the presence of enzymes that degrade cellulose, the principle material of plant cell walls. Large amounts of seagrass are commonly found in the gut contents of this species, likely ingested incidentally while feeding on blue crabs buried among among the seagrass. The presence of these enzymes suggests that Bonnetheads digest the seagrass, which, if true, would be the first instance of omnivory among sharks (planktivory notwithstanding). In 2022, a study suggested Whale Sharks may also be omnivorous, digesting floating algae.

In 2021, Bonnetheads were shown in a series of lab studies to use the Earth's magnetic field for homeward orientation in their annual migrations. Possession of an internal GPS explains the species' site fidelity, or use of the same estuaries annually, after migrations of hundreds of miles.

Bonnetheads were the first sharks in which parthenogenesis, or virgin birth, was observed. A female maintained in a public aquarium for three years with only other female Bonnetheads gave birth to a single healthy female pup. Sperm storage, which is not uncommon in sharks, has never been reported for the species, and moreover, these aquarium specimens were very young juveniles when captured. Subsequently, parthenogenesis in aquarium specimens of sharks and batoids has been documented for numerous species, including the Swell Shark (*Cephaloscyllium ventriosum*; page 42), Whitetip Reef Shark (*Triaenodon obesus*; page 114), Blacktip Shark, Zebra Shark (*Stegostoma tigrinum*), Whitespotted Bamboo Shark (*Chiloscyllium plagiosum*; page 238), and Spotted Eagle Ray (*Aetobatus narinari*). Could this form of reproduction be an adaptation in sharks and their relatives to being isolated from the opposite sex?

→ A Bonnethead, the smallest species of hammerhead (Family Shpyrnidae) swims over seagrass and soft coral communities in Biscayne Bay, Florida.

SHARKS OF THE CONTINENTAL SHELVES

SHARKS OF THE CONTINENTAL SHELVES

Shark habitats on the continental shelves

The Indo-Pacific's Coral Triangle, an area of continental shelf covering about 2.2 million square miles (5.7 million km²) and stretching across the waters of six countries, is the world's preeminent marine biodiversity hotspot, with 3,000 different kinds of fishes. Approximately 30 percent of all named shark and ray species call this area home, and it includes mangrove, seagrass, and other soft-bottom habitats. Sharks found on the continental shelves of other tropical areas, as well as of temperate and polar ecosystems, are an equally fascinating and, in many cases, biodiverse assemblage. In all of these places sharks face challenges from climate change, overfishing, and other human impacts.

SHELF FACTS AND FIGURES

According to the CIA's World Factbook, there are about 221,000 miles (356,000 km) of coastline in the world. Thirty-eight percent, or more than 84,000 miles (135,000 km), border the Pacific Ocean, followed by the Atlantic, Indian, Arctic, and Southern Oceans. Among the countries with the longest coastline, Indonesia ranks first (34,000 miles/55,000 km), followed by Greenland, Russia, the Philippines, and Japan.

↗ A trio of Silky Sharks (*Carcharhinus falciformis*) off Jardins de la Reina, Cuba. Named for the appearance and texture afforded by their small, densely packed dermal denticles, Silky Sharks are among the most wide ranging and common oceanic sharks.

← A Caribbean Reef Shark (*Carcharhinus perezi*) patrols a reef dominated by soft corals and filled with potential prey such as snapper and grunts, Jardins de la Reina, Cuba.

Coastlines are all part of the Earth's continental shelf, the seaward extension of each continent's crustal bedrock that is covered by sediment and ocean. The width of the continental shelf averages about 50 miles (80 km), but it can range from less than 0.6 miles (1 km) to 930 miles (1,500 km), depending on the geological forces that shaped it. Continental shelves cover about 7 percent of the Earth's surface, or about 10.5 million square miles (27 million km^2) and are shallow—usually less than 330 ft (100 m) deep. Much of Earth's continental shelves were dry land during the last glacial period (around 18,000 years ago), when the sea level was much lower, but now virtually all of this area is shark habitat.

SHARK HABITATS OF THE CONTINENTAL SHELVES

A wide variety of sharks live in continental shelf waters, from the tropics to the poles. These habitats are shallow and not too distant from continental land masses. Sharks of the continental shelves are a diverse group with adaptations to their specific environmental conditions, particularly temperature. Tropical continental shelf habitats include coral reefs, mangrove systems, and seagrass beds, the last of which are also found in temperate shelf waters. In temperate latitudes, continental shelf habitats include kelp forests, live bottoms, and other shallow-water, subtidal environments as well as rocky, sandy, and muddy intertidal zones. High latitude continental shelf habitats are mainly polar seas.

Sharks of tropical coasts

Sharks of the tropical continental shelves include many fascinating, ecologically important, and economically significant species, ranging from slender benthic species such as the Whitetip Reef Shark (*Triaenodon obesus*; page 114) to the more robust Tiger Shark (*Galeocerdo cuvier*; page 232). Most are mesopredators, whereas others are part-time top predators.

Tropical ecosystems of the continental shelves occur north and south of the equator where the minimum water temperature does not drop below 68°F (20°C), roughly corresponding to latitudes between the Tropic of Cancer (23.5°N) and Tropic of Capricorn (23.5°S). This expansive area includes the biogeographic realms of the Eastern and Western Tropical Atlantic, Western Indo-Pacific, Central Indo-Pacific, Eastern Indo-Pacific, and Tropical Eastern Pacific, and ecosystems ranging from coral reefs to mangrove forests and seagrass beds. Let's look more closely at some of these tropical marine ecosystems and the sharks that inhabit them.

CORAL REEFS

Coral reefs represent some of the most important real estate on the planet. They cover only about 0.2 percent of ocean floor or 96,000 square miles (250,000 km^2), an area roughly the size of the state of Texas. The major sharks on Indo-Pacific coral reefs include the Grey Reef Shark (*Carcharhinus amblyrhynchos*), Whitetip Reef Shark, Blacktip Reef Shark (*C. melanopterus*), Tiger Shark, Silvertip Shark (*C. albimarginatus*), Galapagos Shark (*C. galapagensis*),

← Grey Reef Sharks (*Carcharhinus amblyrhynchos*) and Blacktip Reef Sharks (*C. melanopterus*) coexist as predators on coral reef systems by occupying slightly different trophic niches, exploiting as prey different parts of the fish and invertebrate community.

SHARKS OF TROPICAL COASTS

A RICHNESS OF SHARKS

More than 200 shark species and 250 kinds of batoids (skates and rays) inhabit shallow tropical ecosystems. About half of these are found exclusively in the tropics. Among the tropical sharks, the most speciose (species-rich) taxa are the cat sharks (family Scyliorhinidae), the houndsharks (Triakidae), and the requiem sharks (Carcharhinidae).

Sharks representative of Indo-Pacific (left) and Atlantic coral reefs

Indo-Pacific coral reefs: Tiger Shark, Great Hammerhead, Sicklefin Lemon Shark, Grey Reef Shark, Silvertip Shark, Galapagos Shark, Scalloped Hammerhead, Blacktip Shark, Blacktip Reef Shark, Whitetip Reef Shark, Ornate Wobbegong, Tawny Nurse Shark and Epaulette Shark.

Atlantic coral reefs: Tiger Shark, Great Hammerhead, Lemon Shark, Caribbean Reef Shark, Blacktip Shark, Bull Shark, Blacknose Shark, Atlantic Sharpnose Shark, and Nurse Shark.

INDO-PACIFIC REEF

ATLANTIC REEF

Blacktip Shark (*C. limbatus*), Scalloped Hammerhead (*Sphyrna lewini*), Great Hammerhead (*Sphyrna mokarran*; page 44), wobbegongs, Tawny Nurse Shark (*Nebrius ferrugineus*), and Halmahera Epaulette Shark (*Hemiscyllium halmahera*; page 80). Sharks on Atlantic reefs include the Caribbean Reef Shark (*Carcharhinus perezi*; page 234), Tiger Shark, Blacktip Shark, Blacknose Shark (*C. acronotus*), Bull Shark (*C. leucas*; page 196), Atlantic Sharpnose Shark (*Rhizoprionodon terraenovae*; page 200), Lemon Shark (*Negaprion brevirostris*; page 230), Nurse Shark (*Ginglymostoma cirratum*; page 40), and Great Hammerhead.

Coral reefs are extremely productive, structurally complex, and biodiverse ecosystems and thus offer a wide variety of fish, invertebrate, and even reptile prey for sharks. Some sharks occupy coral reefs or nearby waters more or less continuously, whereas others are transient visitors. On Indo-Pacific reefs, the former group includes the Grey Reef Shark, Whitetip Reef Shark, and Blacktip Reef Shark, whereas the Tiger Shark, Silvertip Shark, and other species with much larger home ranges (excluding migrations) may pass over and around coral reefs. On coral reefs of the Atlantic, the Caribbean Reef Shark, Blacknose Shark, and Nurse Shark are longer-term occupants.

TROPHIC ECOLOGY ON CORAL REEFS

The prevailing assumption that sharks of coral reefs are apex predators—especially large species such as the Grey Reef and Caribbean Reef Shark—is at best an oversimplification. A 2016 study of the trophic ecology of Australian reef sharks concluded that the Grey Reef, Blacktip Reef, Tawny Nurse, and Whitetip Reef Sharks here are all mesopredators, as are the large bony

↑ Whitetip Reef Shark (*Triaenodon obesus*), a mesopredatory shark common on Indo-Pacific coral reefs, patrols a cave near Gato Island, Philippines.

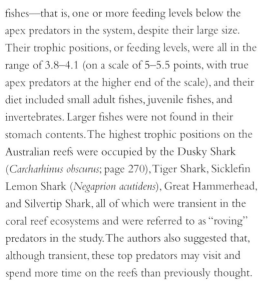

← Whereas the tropical Indo-Pacific supports a diverse assemblage of carpet sharks (Orectolobiformes), the Nurse Shark (*Ginglymostoma cirratum*) is the only benthic representative of the group in the tropical Atlantic Ocean.

→ A Great Hammerhead (*Sphyrna mokarran*), among the top predators in tropical marine ecosystems, breaks the surface in The Bahamas, exposing its tall, distinctive first dorsal fin.

fishes—that is, one or more feeding levels below the apex predators in the system, despite their large size. Their trophic positions, or feeding levels, were all in the range of 3.8–4.1 (on a scale of 5–5.5 points, with true apex predators at the higher end of the scale), and their diet included small adult fishes, juvenile fishes, and invertebrates. Larger fishes were not found in their stomach contents. The highest trophic positions on the Australian reefs were occupied by the Dusky Shark (*Carcharhinus obscurus*; page 270), Tiger Shark, Sicklefin Lemon Shark (*Negaprion acutidens*), Great Hammerhead, and Silvertip Shark, all of which were transient in the coral reef ecosystems and were referred to as "roving" predators in the study. The authors also suggested that, although transient, these top predators may visit and spend more time on the reefs than previously thought.

In addition, they recommended adding a descriptor to the trophic designation of the coral reef sharks that do not meet the definition of true apex predators: "high-level" mesopredators.

SHARK LAB

The Bimini Biological Field Station, more commonly called the Shark Lab, is on the westernmost of the islands of The Bahamas, between the deep Gulf Stream and shallow Great Bahama Bank. Established by legendary shark biologist Samuel "Sonny" Gruber, the lab has conducted the world's most extensive research on the ecology, behavior, life history, and physiology of the Lemon Shark, with additional contributions on other sharks, including Tiger Sharks, Bull Sharks, Great Hammerheads, and Caribbean Reef Sharks.

SHARKS OF THE CONTINENTAL SHELVES

↑ Two sleek Caribbean Reef Sharks push through a school of Blue Runners (*Caranx chrysos*).

↗ After being tagged and released by researchers, a large Tiger Shark (*Galeocerdo cuvier*) swims across the crystal clear waters of the Great Bahama Bank near Bimini.

SHARKS OF TROPICAL COASTS

SHARKS OF THE CONTINENTAL SHELVES

> ### HOW DO YOU USE YOUR ENERGY?
>
> The energy provided by its prey is apportioned to varied uses in the Lemon Shark's energy budget. About 22.4 percent of the calories a 2.2 lb (1 kg) juvenile Lemon Shark eats are used for growth; 49.4 percent for its metabolism, including energy for swimming, ventilating gills, and body maintenance; and 28.2 percent is lost in urine and feces. Since energy transformations are inefficient, significant energy is lost as heat in the overall process. Thus, 1 ounce (28 g) of prey does not build 1 ounce (28 g) of Lemon Shark.

WHEN LIFE GIVES YOU LEMONS

From birth to about three years in age and a length of about 3 ft (1 m), the Lemon Sharks at Bimini occupy a narrow strip of mangrove-lined nurseries. At ages ranging from five to eight years, female Lemon Sharks migrate from Bimini to as far north as Virginia. At maturity, at about age 14–17 years, the females mate and return for the first time to where they were born to have their own pups, a phenomenon called natal philopatry, the roots of which mean "birth" and "home-loving." Subsequently, the females return annually or every other year to pup, an activity common to a diverse array of shark species, including the Bonnethead (*Sphyrna tiburo*; page 204), Sandbar Shark (*Carcharhinus plumbeus*; page 198), and Sand Tiger (*Carcharias taurus*; page 236). The method by which these sharks migrate remained only hypothetical until recently, when one of our graduate students demonstrated that Bonnetheads use the Earth's magnetic field to determine their position and to navigate, although other mechanisms may be involved.

→ A newborn Lemon Shark (*Negaprion brevirostris*) swims among the underwater prop roots of a Red Mangrove (*Rhizophora mangle*). These prop roots provide critical cover and protection to the young sharks.

SHARKS OF TROPICAL COASTS

SHARKS OF THE CONTINENTAL SHELVES

Juvenile Lemon Sharks in Bimini were the subject of the first experimental study of a displaced shark's ability to return to its home. In the study, 32 Lemon Sharks aged two years old or younger were individually captured in their nursery in Bimini, equipped with ultrasonic transmitters (pingers), and relocated about 2.5–10 miles (4–16 km) away. Once an individual was released, it was tracked from a small boat using a directional hydrophone that detected the unique signal of its pinger. Thirty-one of the 32 sharks successfully returned to Bimini, 26 of them to the specific, small, well-defined home ranges from which they were captured. Miraculously, one shark that had been displaced in the Gulf Stream, a powerful warm current with depths exceeding 2,600 ft (900m), managed to find its way back home! According to the researchers, cues that could have played a role in the innate ability

↑ A group of Lemon Sharks jostle for position on a shallow sand flat, likely competing for common prey such as Yellowfin Mojarra (*Gerres cinereus*) and Blue Crabs (*Callinectes sapidus*).

← Researchers at the Bimini Biological Field Station tag a neonate Lemon Shark as part of an extensive work-up before releasing it. The ecology, behavior, life history, and physiology of Lemon Sharks have been extensively studied for over 30 years by Dr. Samuel Gruber and his graduate students and associates at the site.

of the young Lemon Sharks to return to their home, or nearby, include local water currents, sensory cues (such as smell and even sound), gradients of water depth, and the geomagnetic signature of their home habitat.

Studies of the behavior of Lemon Sharks have also yielded interesting findings about their preference for familiars ("friends") over strangers, their personalities, and their ability to learn. One study showed that juvenile Lemon Sharks preferred other Lemon Sharks to similar-sized Nurse Sharks, presumably an adaptation to form groups that might protect them from predators venturing into their habitat at high tide, likely through the increased sensory ability of a group over an individual. Juvenile Lemon Sharks were also most likely to socialize with other Lemon Sharks of similar sizes, and even had preferences for familiar individuals over strangers. In addition, juvenile Lemon Sharks that had been trained to touch a target— in itself a noteworthy accomplishment—facilitated learning in sharks that had not been previously trained when paired with them. Educational psychologists, take note!

One final point about Lemon Sharks: their diet consists mainly of fish (including small sharks in the case of adult Lemon Sharks), crustaceans, and mollusks. But how much do they eat a day, and how is that food utilized? Studies by Gruber have shown that a juvenile Lemon Shark up to nearly 7 ft (2 m) long consumes 1.5–2.1 percent of its body weight a day, a value known as its daily ration. For a 2.2 lb (1 kg) Lemon Shark, the daily ration would therefore be 0.53–0.74 oz (15–21 g). A 3 inch-long (7.5 cm) Spotfin Mojarra (*Eucinostomus argenteus*), a favorite prey item, weighs less than 1 oz (28 g). An adult Lemon Shark would consume about 3.5 lb (1.5 kg) of larger fish (including juvenile Lemon Sharks) daily to meet its nutritional needs.

SHARKS OF THE CONTINENTAL SHELVES

Mangrove systems and seagrass beds

Mangroves are tropical trees that can tolerate salt and grow at the water's edge between latitudes 32°N and 38°S. Seagrass beds are diverse systems found in shallow waters that are home to submerged aquatic vegetation, primarily marine flowering plants such as Turtlegrass (*Thalassia testudinum*) and eelgrasses.

Mangrove, seagrass, and coral reef communities are sometimes cited as examples of linked habitats because of their close association and importance to one another. Like coral reefs, mangrove systems and seagrass beds represent habitat for a variety of sharks, including Lemon Sharks, Sicklefin Lemon Sharks, Nurse Sharks, Tawny Nurse Sharks, Tiger Sharks, Bull Sharks, Blacktip Sharks, and Bonnetheads. Smalltooth Sawfish (*Pristis pectinata*; page 112) also use these habitats.

AN AUSTRALIAN EXCEPTION

While mangrove and seagrass ecosystems are used by juvenile Lemon Sharks for refuge and foraging (pages 71 and 180), they are often too shallow for larger species of sharks. In addition, the increased temperature—especially of exposed seagrass beds—may impose a physiological thermal stress that large species or larger individuals of some species cannot tolerate. A major exception to this is the Tiger Shark (*Galeocerdo cuvier*; page 232), the most common large shark of the shallow seagrass ecosystem of Shark Bay, Western Australia. In that system, occurrence of Tiger Sharks is positively correlated with water temperature: the higher the temperature, the more frequent their occurrence. However, the higher temperatures are not the direct cause of the increased presence of Tiger Sharks in Shark Bay. Instead, it is the increased prey density—particularly sea snakes and Dugongs (*Dugong dugon*), but also sea turtles and smaller sharks and rays—that attracts them. Since the Tiger Sharks' prey in this system are either herbivores or low-level carnivores, they are considered mesopredators in Shark Bay, despite their large size and reputation as ultimate predators. Although their prey may remain there in the winter, Tiger Sharks migrate away when water temperatures cool.

↑ A juvenile Smalltooth Sawfish (*Pristis pectinata*) swims along a Turtlegrass (*Thalassia testudinum*) bed. Mangrove habitats with adjacent shallow seagrass beds are critically important for Smalltooth Sawfish during their first year of life.

← A juvenile Blacktip Shark (*Carcharhinus limbatus*) cruises a sand flat along a Red Mangrove fringed shoreline.

SHARKS OF TEMPERATE COASTS

Nearshore ecosystems of temperate coasts include shallow-water subtidal environments (including kelp forests and live bottoms) and intertidal zones (rocky, sandy, and muddy). The former is habitat for numerous species of sharks, including White Sharks (*Carcharodon carcharias*; page 106), whereas the intertidal zone's harsh environmental conditions, including its high-energy wave action, limit the number of species.

Many of the shark species found in estuaries (page 184) also occupy temperate shallow-water ecosystems, such as the Sandbar, Bull, Lemon, and Port Jackson (*Heterodontus portusjacksoni*; page 76) Sharks, as well as the Bonnethead and Spiny Dogfish (*Squalus acanthias*; page 264). Other well-known temperate coastal sharks include the Tiger Shark, Sand Tiger, Finetooth Shark (*Carcharhinus isodon*), White Shark, Dusky Shark, Dusky Smoothhound (*Mustelus canis*; page 266), Galapagos Shark, Leopard Shark (*Triakis semifasciata*), Angel Shark (*Squatina squatina*; page 272), Spinner Shark (*Carcharhinus brevipinna*), Small-spotted Catshark (*Scyliorhinus canicula*), and Broadnose Sevengill Shark (*Notorynchus cepedianus*).

HAPPY EDDIE INDEED

Another denizen of the southern Africa coastline is the Puffadder Shyshark (*Haploblepharus edwardsii*), also known as the Happy Eddie Shyshark. A member of the large cat shark family, it inhabits shallow rocky and sandy areas. "Puffadder" refers to the shark's coloration, which is similar to that of the Puff Adder snake (*Bitis arietans*). Its alternative common name, Happy Eddie, is a diminutive of the scientific name. And "shyshark" derives from the species' habit of rolling into a ball with its tail covering its eyes when it feels threatened, such as by its predators the Broadnose Sevengill Shark and Cape Fur Seal (*Arctocephalus pusillus*), at which times its happiness is dubious. It is also called the Donut Shark and, if it were found in New York, we suspect it would be known as the Bagel Shark.

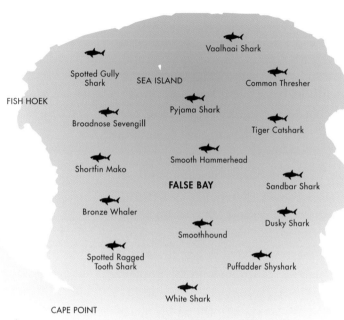

False Bay, South Africa
At least 15 and as many as 22 species of sharks are reported from False Bay, a high-salinity 420-square-mile (1,090 km²) water body near Cape Town, South Africa. Once a hotspot for juvenile and sub-adult White Sharks (*Carcharodon carcharias*), their numbers have declined significantly since 2016.

TEMPERATE HOTSPOT

One hotspot for temperate-water sharks is the coast of southern Africa. Two species of this area are the Pyjama Shark (*Poroderma africanum*, also known as the Striped Catshark) and the Broadnose Sevengill Shark. The Pyjama Shark, a bottom-dwelling member of the cat shark family endemic to the region, is perhaps best known for its villainous role stalking the protagonist in the 2020 Netflix documentary *My Octopus Teacher*. It is also the only shark with head-to-tail horizontal stripes. It can be found in shallow-water nearshore environments, including kelp beds.

The Broadnose Sevengill Shark may exceed 9 ft (2.7 m) in length and is among the most primitive sharks swimming today. In the kelp forests of False Bay, near Cape Town, these sharks eat other sharks, Cape Fur Seals (*Arctocephalus pusillus*), bony fishes, and some invertebrates. One study determined that the Broadnose Sevengill Sharks in False Bay occupy a higher trophic level than the White Sharks inhabiting the same ecosystem, in part because the sevengills ate more seals than the White Sharks. In the late 2010s, Broadnose Sevengills expanded their range in False Bay, occupying habitat vacated by White Sharks, which had moved to avoid predation by Orcas (*Orcinus orca*) that had recently entered the bay.

SHARKS OF THE CONTINENTAL SHELVES

Sharks of polar coasts

It's cold in polar seas, but that doesn't stop sharks from living there. A small assemblage of sharks occupies the Arctic Ocean, including the massive Greenland Shark (*Somniosus microcephalus*; page 202) and Pacific Sleeper Shark (*S. pacificus*), as well as the Porbeagle (*Lamna nasus*; page 110) and Basking Shark (*Cetorhinus maximus*; page 46). The southern hemisphere has fewer polar sharks, which include the Southern Sleeper Shark (*Somniosus antarcticus*) and Porbeagle. No shark species lives exclusively in polar shelf waters.

Temperature is one of the most important environmental factors in the life of sharks, and indeed all organisms. It affects not only the diversity, abundance, and distribution of organisms, but also an individual's behavior, activity, growth, development, metabolism, heart rate, digestive rate, and blood characteristics. Temperature affects all levels of the organism's physiology, from the molecular to whole organ systems.

→ A Salmon Shark (*Lamna ditropis*) off Alaska. Found only in the North Pacific Ocean, the Salmon Shark, named for one of its primary prey sources, is among the few coastal or pelagic shark species that can tolerate cold subpolar temperatures.

CHILLING OUT

Sharks have both thermal preferences and upper and lower thermal limits, and these vary not only by species, but also in populations of the same species that live in different locations, such as different latitudes or depths with different temperature regimes, and by life stage and sex. Within the range of temperatures that sharks are found in, and with all other variables (such as prey availability) being equal, there are advantages to living nearer higher temperatures. These include faster digestion, processing of sensory information, muscle contractions, and rates of chemical reactions, along with faster development in utero and overall growth rate. But there are opportunities for sharks living in cold-water environments such as the polar seas, including decreased competition and therefore increased availability of suitable prey.

SHARKS OF THE CONTINENTAL SHELVES

Sharks occupy both the Arctic and Antarctic polar regions, broadly defined as those areas whose latitudes exceed about 66°N or 66°S, corresponding to the Arctic and Antarctic Circles, respectively. The major bodies of water are the Arctic Ocean to the north and the Southern Ocean to the south. Winter sea-surface temperatures at these locations remain constantly cold—as low as the freezing point of seawater (28.5°F, -1.9°C) to about 35.5°F (2.0°C).

The biodiversity of polar sharks is low, with only about a dozen species in Arctic waters and three in the Antarctic, all of which also occur in the cold-temperate zones. This trend of decreasing biodiversity at higher latitudes applies across all taxa, not just sharks.

Buoyancy in sharks

Sharks are heavier than the seawater in which they reside and, unless they rest on the bottom, they require additional lift. Dynamic lift is generated by the fins during swimming. Static lift is provided by their oily livers, among the biggest of which are found in Greenland Sharks (*Somniosus microcephalus*) (although we once dissected a large White Shark (*Carcharodon carcharias*) and its liver was enormous as well). The cold water in which the Greenland Shark lives is dense and also contributes to buoyancy in the species.

A SLUGGISH HULK

Let's consider the Greenland Shark. It is in the sleeper shark family (Somniosidae; *somni* is Latin for "sleep"), and documentary videos depict a sluggish hulk of a beast moving at about the speed of a lethargic manatee. Cruising speeds of Greenland Sharks are indeed slow: one estimate was 0.5 mph (0.8 kph), which is similar to that of a Sand Tiger (*Carcharias taurus*; page 236) and about a quarter that of a cruising White Shark. While the Sand Tiger swallows air to achieve near-neutral buoyancy (page 236), the Greenland Shark's neutral buoyancy is due to its large, oily liver and the supportive lift of its denser cold-water environment. However, the stomach contents of the species tell a different story of its swimming speed. The Greenland Shark eats marine mammals such as seals and whales, as well as squid and fish, indicating that it is capable of bursts of speed sufficient to catch such highly mobile, speedy prey. Why doesn't the Greenland Shark—the largest

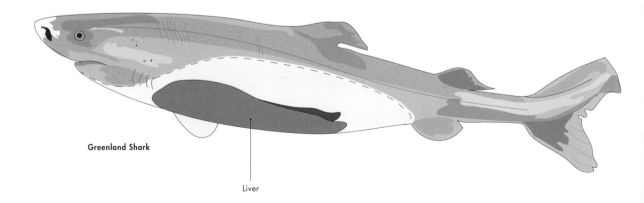

Greenland Shark

Liver

polar fish—freeze in its frigid environment? Apparently, it uses a variation of the method employed by transportation bodies and residents of areas prone to freezing: adding salt to lower the freezing point. In the case of the Greenland Shark, the "salt" is trimethylamine N-oxide (TMAO), which is similarly thought to depress the freezing point. TMAO also makes the flesh distasteful.

Greenland Sharks are found in high-latitude fjords and have also been caught deep in the Gulf of Mexico, in water temperatures similar to those in their polar homes. Since the species is physiologically restricted to cold-water environments, rising temperatures in the polar seas caused by global warming may have a disproportionate impact on it, as well as on other polar marine organisms.

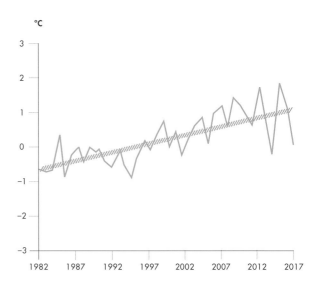

↑ Seawater temperatures in the Arctic may fluctuate annually (shown here by the solid line), but the trend (the slashed line) clearly shows that these waters are warming, and doing so faster than in temperate and tropical oceans. These changes are largely a result of human-caused climate change, and the impacts on the sharks and other marine life of the area may be dramatic in the long term.

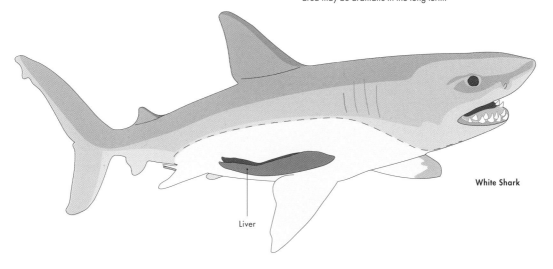

Liver

White Shark

SHARKS OF THE CONTINENTAL SHELVES

NEGAPRION BREVIROSTRIS

Lemon Shark

There's no place like home for Lemon Sharks

SCIENTIFIC NAME	*Negaprion brevirostris*
FAMILY	Carcharhinidae
NOTABLE FEATURE	Two dorsal fins of approximately the same size, muted lemon coloration, non-serrated teeth, blunt snout
LENGTH	11 ft (3.4 m)
TROPHIC LEVEL	Predator of sharks, bony fishes, crustaceans, and mollusks

Lemon Sharks were the first species of shark for which sperm storage and multiple paternity (whereby a single litter could have four or more different fathers) were documented. Subsequently, these phenomena have been shown to be more the rule than the exception in sharks. Lemon Sharks are also philopatric, returning to their birthplace to pup.

The scientific name for the Lemon Shark, *Negaprion brevirostris*, translates roughly as "no serrations" and "short snout," in reference to the species' smooth teeth and blunt head. Lemon Sharks inhabit temperate and tropical waters of the Atlantic and coasts of the Tropical Eastern Pacific. Juveniles are born in shallow, mangrove-fringed lagoons, and adults occupy coral reefs, seagrass beds, and other coastal habitats. Mangrove habitats in particular are critical to the growth and survival of juvenile Lemon Sharks during their first years of life, when they are preyed on mainly by larger sharks.

In The Bahamas and Brazil, juvenile Lemon Sharks are known to move with the tides. At high tide they inhabit the upper reaches of the mangrove systems, which offer them a degree of refuge from predators that enter the lagoons at this stage of the tidal cycle. At low tide, juveniles in Bimini in The Bahamas will venture into the relatively safe spaces of adjacent seagrass beds to forage for small fish and invertebrates. Off Cape Canaveral, Florida, juveniles aggregate in a different, more challenging environment—the high-energy surf zone off the sandy beaches, where the shallows offer a significant measure of protection.

Lemon Sharks are known to lie motionless on the bottom. Like other sharks, they use ram ventilation when swimming, keeping their mouth open and thus irrigating their gills with fresh seawater, which both provides oxygen and carries carbon dioxide away. When resting on the seafloor, Lemon Sharks switch to buccal pumping, alternating expanding and compressing the oral cavity to move water over the gills. The IUCN categorizes Lemon Sharks as Vulnerable.

→ Lemon Sharks, particularly juveniles, have been extensively studied and have provided valuable information, some of which can be applied to sharks in general. These include insights into homing, mating, foraging, behavior, vision and other senses, metabolism, reproduction, and ecology.

SHARKS OF THE CONTINENTAL SHELVES

GALEOCERDO CUVIER

Tiger Shark

Apex predator with can-opener teeth

SCIENTIFIC NAME	*Galeocerdo cuvier*
FAMILY	Galeocerdidae
NOTABLE FEATURE	Unique cockscomb-shaped teeth in both jaws, distinctive tiger-like markings on flanks, ridge on dorsal surface anterior to first dorsal fin
LENGTH	16.5 ft (5 m)
TROPHIC LEVEL	Predator of sea turtles, birds, sharks, bony fishes, and invertebrates

As iconic species go, the Tiger Shark rivals the White Shark as the most widely recognized and, unfortunately, feared shark. Juvenile Tiger Sharks feed mostly on bony fishes. As adults, they commonly eat bony fishes, sharks and rays, sea turtles, sea snakes, birds, and mammals, and even invertebrates such as crustaceans and mollusks.

The Tiger Shark has very recently been moved from the requiem shark family (Carcharhinidae) to its own family, Galeocerdidae, of which it is the only living species. The rational for this, validated by genetic analysis, includes its unique reproduction strategy (see below), different jaw structure, unique teeth, presence of a predorsal ridge, homodont dentition (the same tooth type in the upper and lower jaws), and keeled tail.

The mode of embryonic nutrition in Tiger Sharks that separates the species from the carcharhinids is known as embryotrophy. In this, developing embryos are sequestered in sacs containing a clear nutritive fluid called embryotroph, which they absorb as they grow. Carcharhinids have a placental connection that supplies nutrition from the mother to embryo.

Tiger Shark litters may exceed 60 pups, which are born after a gestation period of 15–16 months at around 30 in (75 cm) in length, making them vulnerable to predators. They grow rapidly and reach sexual maturity early, at 7–10 years depending on the location.

One surprising feature of the Tiger Shark is its relatively weakly calcified jaws, similar to those of the cow sharks, which explains why the dried jaws of both the Tiger Shark and cow sharks are usually bent or deformed. Its relatively weak jaws are adapted to bending across the body of large prey (such as sea turtles and dead whales), such that all of its teeth make contact with the flesh. The shark then twists or spins its body to carve out huge chunks of flesh with its heavily serrated teeth.

Tiger Sharks are considered Near Threatened by the IUCN, but populations are stable or increasing in some locations, including the East Coast of the United States and Bimini in The Bahamas.

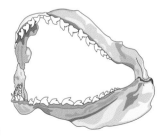

Formidable but flexible
The jaws of Tiger Sharks, especially large specimens, are intimidating, but they are less calcified than those of most other sharks, and bend to allow the specialized teeth to make better contact with the prey, from which it shears chunks of tissue.

→ A large Tiger Shark swimming over seagrass, perhaps in search of sea turtles, among their most common prey.

SHARKS OF THE CONTINENTAL SHELVES

CARCHARHINUS PEREZI
Caribbean Reef Shark
Coral reef shark of the Caribbean

SCIENTIFIC NAME	*Carcharhinus perezi*
FAMILY	Carcharhinidae
NOTABLE FEATURE	Short snout, large first dorsal fin, ridge between dorsal fins (interdorsal ridge), black border on lower lobe of caudal fin
LENGTH	10 ft (3 m)
TROPHIC LEVEL	Predator of bony fishes, sharks, rays, crustaceans, and cephalopods (primarily squid)

In spite of being the most common reef shark of the Caribbean Sea, the Caribbean Reef Shark remains relatively poorly studied, unlike its Indo-Pacific ecological counterpart, the Grey Reef Shark. An economically important ecotourism industry revolves around Caribbean Reef Sharks in The Bahamas, and our students in Bimini are given the privilege of snorkeling among these magnificent predators. The IUCN classes the species as Endangered.

A study in Eleuthera, in The Bahamas, found that Caribbean Reef Sharks spend most of their time on the outer reef and not in the reef's inner reaches. The same study found that males likely had larger home ranges than females, and that larger individuals preferred waters further offshore than smaller sharks. Segregation by sex and size class, as demonstrated by the study, is not unusual among sharks.

A 2018 study of the trophic ecology of Caribbean Reef Sharks found that the species did not act as an apex predator on isolated coral reefs, but rather participated as part of a tag team of large bony fishes such as barracuda and Black Grouper (*Mycteroperca bonaci*) in what the authors called an upper-trophic-level predator guild. Thus, at least in the case of similar small coral reef ecosystems, it would be incorrect to call Caribbean Reef Sharks either apex predators or mesopredators, a predatory nomenclature conundrum of sorts.

In any case, sharing the top predator role among several species apparently serves the reef system well, since it buffers the ecosystem somewhat from the potential loss of a single apex predator.

Although we have mentioned shark ecotourism here, we have not discussed the ecological impact of the practice. A study of provisioning (principally hand-feeding) Bull Sharks has shown that in some sites individuals might meet their entire metabolic requirement by being hand-fed. In such cases, the consequences to the health of the individual sharks and their ecosystem are not yet known. In another study, only a small percentage of Caribbean Reef Sharks at a site where provisioning had occurred for 20 years dined regularly at the buffet, and researchers thus concluded that the impacts of the practice might be limited. In contrast to the Bull Sharks that were studied, the Caribbean Reef Sharks very likely could not meet their metabolic needs exclusively through provisioning at the study site. Provisioning did not affect movements of the sharks either. The study included the cautionary note that its results should not be extrapolated to provisioning in general, and that more research is needed. We agree.

→ The Caribbean Reef Shark is an elegant and ecologically important predator of inshore waters of the Atlantic. Climate change that impacts their habitats, especially coral reefs, is a major threat to the species.

SHARKS OF THE CONTINENTAL SHELVES

CARCHARIAS TAURUS

Sand Tiger
A lethal sibling

SCIENTIFIC NAME	*Carcharias taurus*
FAMILY	Odontaspididae
NOTABLE FEATURE	Flattened head, conical snout, both dorsal fins similar in size, large anal fin, pointed teeth with two smaller lateral cusplets
LENGTH	About 10 ft (3 m)
TROPHIC LEVEL	Generalist fish-eater, including of skates, eels, and sharks

Sand Tigers are commonly displayed at large marine aquariums. In one incident, a group of schoolchildren visiting a newly opened public aquarium were mesmerized by its enormous snaggle-toothed Sand Tiger, which was almost hovering in the main tank. Before their young eyes could drink in what had happened, the Sand Tiger accelerated at what seemed like breakneck speed and swam away with a smaller Blacktip Shark in its mouth, leaving behind a billowing trail of red-stained water. Like the poor Blacktip Shark, the wide-eyed kids had been misled by the Sand Tiger's initial slothful, sluggish movement. Predator–prey interactions like this one play out continuously throughout the animal kingdom as part of the natural struggle for existence.

Sand Tigers are easily recognized by their protruding mouthful of long teeth, seemingly pointing randomly (hence their other common name, Ragged-Tooth Shark), as well as by their slow movement, swimming as if through molasses in their aquarium tank along with the groupers and a few other unhurried bony fishes. The slow movement and hovering of the bony fishes is explained by the presence of an inner balloon, the gas-filled swimbladder, but sharks do not possess such an organ. Instead, the Sand Tiger hangs in place after gulping air at the surface and storing it in its stomach, the lining of which is impervious to air, thus saving on precious energy that it would otherwise need to propel itself and keep from sinking.

Sand Tigers are also known for another adaptation, embryophagy, or intra-uterine cannibalism. Teeth develop earlier in Sand Tiger embryos than in other sharks. When it reaches a size of about 5 in (12 cm) in length, the largest embryo with well-developed teeth in each uterus will consume the others—as many as a dozen of its siblings—and stores the energy in its yolk stomach. Females give birth to two large young (one in each uterus), each more than 3 ft (1 m) long and thus formidable predators with a high survival probability at birth. Sand Tigers are listed by the IUCN as Critically Endangered.

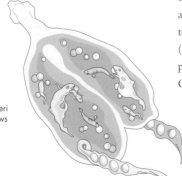

Evolution in brutal form
Embryophagy, or intrauterine cannibalism, is illustrated in the uteri of the Sand Tiger. Lower right shows ovaries containing ova. In each uterus, the largest shark with its precocious teeth uses its sibling roommates as food.

→ The Sand Tiger, also called the Ragged Tooth Shark or Grey Nurse Shark, glides effortlessly under a ledge at Aliwal Shoal, South Africa.

SHARKS OF THE CONTINENTAL SHELVES

CHILOSCYLLIUM PLAGIOSUM

Whitespotted Bamboo Shark

Egg-laying carpet shark

SCIENTIFIC NAME	*Chiloscyllium plagiosum*
FAMILY	Hemiscylliidae
NOTABLE FEATURE	Long tail, dark bands covering brown body, white and black spots, lobed fins
LENGTH	Less than 3 ft (1 m)
TROPHIC LEVEL	Small bony fishes and invertebrates

The Whitespotted Bamboo Shark is a sluggish, benthic, nocturnally feeding shark of Indo-Pacific coral reefs. It is widely maintained in marine aquariums, even by hobbyists, and can reproduce successfully in captivity, which has led to numerous studies of its biology.

The Whitespotted Bamboo Shark is one of the approximately 40 percent of sharks and rays that are oviparous, or egg-layers. It will lay about two dozen eggs annually, one or two at a time over a two-month period, and the eggs take about 3–4 months to hatch. The source of nutrition for the developing embryo throughout the entire developmental period is the yolk deposited by the mother before the eggshell is formed, after which there is no additional maternal contribution to embryonic nutrition. As the embryonic Whitespotted Bamboo Shark develops, it uses up the yolk to supply energy for growth, respiration, and metabolism, and its birth weight is 20–25 percent less than its weight in the early stages. The embryo defecates and urinates within the egg case, and if the internal environment of the egg was completely sealed off, the accumulation of these waste products would soon kill it. To avoid this, there are openings in the egg case to the surrounding seawater, and movement of the embryo facilitates the exchange of fluids between the interior and exterior.

Developing shark in its egg case
The case is tethered by tendrils to a sponge, with the eggs hatching after 14–15 weeks. The hatchlings are about 6 in (15 cm), and resemble miniature adults. As in all sharks, there is no parental care.

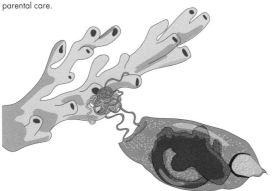

→ The Whitespotted Bamboo Shark, commonly exhibited in aquaria, is considered Near Threatened because it is being overfished for food in the Indo-Pacific.

SHARKS AND US

SHARKS AND US

How we help sharks

Let's begin this chapter by celebrating the enormous reservoir of respect and even veneration with which sharks are viewed by large numbers of people. A seemingly endless stream of television documentaries about sharks—not all on gory attacks—attracts huge audiences, and social media posts of striking photos and videos of sharks abound. Across the globe, classes on shark biology and conservation are springing up on college campuses and online. Governments are creating shark sanctuaries, and fishery managers are effectively reversing trends of declining populations in numerous species. Even economists are concluding that the financial value of healthy sharks in their habitats exceeds their value as food or products.

→ A Lemon Shark (*Negaprion brevirostris*; left) and a Caribbean Reef Shark (*Carcharhinus perezi*; right), among the most common large coastal shark species throughout the tropical western Atlantic Ocean.

SHARKS DO NOT EXIST IN A VACUUM

To ensure shark populations are healthy, the ecosystems in which they reside also need to be healthy, with the appropriate mix of organisms at different trophic levels, as well as the required environmental conditions to support all of the ecosystem's inhabitants. Conservation based on this concept, aptly called ecosystem management, is complex, in part because effective management requires an understanding of the demographics, life history characteristics, and health of all the species in the ecosystem, not only its sharks, as well of the human impacts on the ecosystem.

OBJECTS OF FASCINATION, BUT…

People and governments are starting to value sharks for their ecological roles, their economic importance, and their intrinsic worth, or right to exist without further justification. For most, our fascination with sharks derives from the adaptations they possess that make them such effective predators, that enable them to avoid being prey themselves, and that they use when mating, all of which we have highlighted throughout this book. For others, like commercial fishers and ecotourism operators, sharks are central to their livelihoods. And, let's just say it (and our illustrations prove it): sharks are stunningly beautiful, graceful, awe-inspiring beasts.

However, we must also acknowledge that the actions of humans threaten the existence of sharks, and for a variety of reasons. First, the life history characteristics of most sharks are not sufficiently understood to allow us to conserve and manage them effectively, especially those species living in the deep sea. To manage sharks, biologists need to undertake assessments of the current populations for every species, and they should know their diet, fecundity, gestation period, life span, growth rate, age at maturity, size at birth, migration patterns, and so on. This basic life history information is unknown for nearly all of the angel sharks (family Squatinidae), yet at least 12 species are currently listed on the IUCN Red List as Endangered or Critically Endangered. Acquiring these data takes time, money, and other resources, not to mention trained shark scientists.

Second, in most cases sharks have slow (or conservative) life history characteristics—that is, few offspring after long gestation periods, slow growth rates, and late maturity. For example, Dusky Sharks

(*Carcharhinus obscurus*; page 270) typically take more than two decades to reach maturity and produce only 8–12 pups every three years because their gestation period is nearly two years. Third, shark scientists do not sufficiently understand the health, behavioral, and ecological impacts of many environmental threats on sharks. And finally, overfishing—legal and not—of many shark species continues. There is one piece of good news: although many shark populations have experienced significant decreases due to overfishing and other human insults, as far as we know humans have not driven any shark or ray species to extinction in the Common Era.

↑ A Great Hammerhead (*Sphyrna mokarran*) comes straight toward the camera at a shark ecotourism dive site in Bimini, Bahamas.

↖ As the largest fish in the sea, but one that feeds on plankton, swimming with Whale Sharks (*Rhincodon typus*) has become a popular and economically valuable form of ecotourism.

SHARKS AND US

↑ The tiny eyes but very large chemical sensing barbels of the Nurse Shark (*Ginglymostoma cirratum*) pretell the primary sensory system this benthic shark uses to locate its often hidden prey.

↗ Multiple species of sharks including Silky Sharks (*Carcharhinus falciformis*), Galapagos Sharks (*Carcharhinus galapagensis*), and Blacktip Sharks (*Carcharhinus limbatus*) work along with Yellowfin Tuna (*Thunnus albacares*) to shoal bait into a bait ball as they take turns attacking their prey.

HOW WE HELP SHARKS

SHARKS AND US

How we hurt sharks

Sharks live in oceans imperiled by a suite of anthropogenic impacts. These range from overfishing, bycatch, pollution, and habitat destruction to the far-reaching effects of global climate change.

Human-caused climate change warms and acidifies the environments in which sharks live, and touches virtually every aspect of their (and every other organism's) biology and ecology. Runoff from industrialized agriculture and undertreated human sewage create both temporary and permanent dead zones—one monitored in 2017 in the Gulf of Mexico exceeded 8,700 square miles (22,500 km^2) and had oxygen levels too low to sustain animal life, including sharks. Micro- and macro-plastic pollution is burgeoning, as are other types of pollution, including other chemicals, noise, light, and radioactivity. To add to these are the threats of habitat destruction, loss of biodiversity in ecosystems that sharks inhabit, and overfishing. And just when you thought it could get no worse, a previously underrecognized impact of humans enters the picture—"roadkill" (see box).

← A dead requiem shark in a gill net. Once the shark's head penetrates through the mesh, it can neither move forward nor reverse itself, and will struggle before dying. Gill nets may continue to indiscriminately catch sharks and other organisms even after they are abandoned.

ROADKILL

Many of us routinely espy dead animals on our roads, the victims of the intersection of fast, heavy vehicles and naive mammals, birds, reptiles, amphibians, and insects, whose habitats have been invaded by human development and divided by highways. But few people realize that large sharks are also "roadkill" victims, especially the slow-moving planktivorous species of the ocean's surface such as the Basking Shark (*Cetorhinus maximus*; page 46). Obviously not seen on our terrestrial roads, these casualties are found along oceanic thoroughfares. And the phenomenon is not new: a 2008 Indian Ocean report included statistics for Whale Shark (*Rhincodon typus*; page 144) collisions with ship hulls and propellers. In addition to being slow-moving, Whale Sharks and Basking Sharks are known to make long migrations, some of which take them across busy shipping channels. A 2022 study found a higher incidence of ship collision with Whale Sharks in gulf regions where high traffic coincided with Whale Shark migratory pathways.

Overfishing

Shark fishing in United States waters accelerated in the 1980s, when stocks of mainstay bony fishes declined due to overharvesting, and international trade relationships opened with China, a change that occurred globally. That's when shark fins—the essential ingredient in shark fin soup, a traditional delicacy and status symbol in countries in East and Southeast Asia—became the most economically valuable part of these target species.

SHARK FINS

There is a distinction between finning—which is illegal in many countries, including the United States—and the legal shark fin trade. In finning, the fins are cut from the shark, most often immediately after capture when the animal is still alive. The finless, dying shark, which has much less commercial value than its fins, is then unceremoniously heaved back into the water where it slowly dies. If a shark is caught legally and the meat brought to market, then the fins can be legitimately sold. Bans or restrictions on trading illegal fins have been enacted in numerous countries around the world. In the U.S. the Shark Fin Sales Elimination Act is under consideration in the US, which would ban sales of all shark fins regardless of whether the fishery is sustainable.

Combined, shark and ray fisheries comprise less than 1 percent of the total marine capture fishery, but that does not mean that overfishing of sharks is not occurring. Existing at or near the top of the food chain, shark populations are necessarily small compared to the bony fishes that are often targeted by large fisheries and live lower on the food chain. Thus, even low harvest rates of large sharks, in comparison to other fishes, can lead to overfishing. The top five countries with the highest landings of sharks and rays are India, Indonesia, Mexico, Spain, and Taiwan. These five, plus Argentina, the United States, Pakistan, Malaysia, and Japan, make up about 60 percent of the shark and ray landings worldwide. The remaining 40 percent are small countries, mostly island nations or poorer countries in Africa. These are mostly small-scale, artisanal fisheries, from which good data are difficult to obtain.

OTHER SHARK PRODUCTS

In addition to shark fins, economically valuable products from sharks include shark meat, leather from shark skin, cartilage for the alternative healthcare industry, livers for omega-3 fatty acids, liver oil as an ingredient in cosmetics, and teeth and jaws. Before it is eaten, some shark meat requires soaking or marinating to remove both the urea that occurs naturally in all shark tissue and the extremely distasteful compound ammonia, which forms in dead sharks as the urea degrades. If you must eat sharks, make sure you consume only those species that are sustainably fished—for example, Dusky Smoothhounds (*Mustelus canis*; page 266) and Blacktip Sharks (*Carcharhinus limbatus*) on the East Coast of the United States, and Gummy Sharks (*Mustelus antarcticus*) in southern Australia—and are from well-managed stocks, and that you follow state and federal laws when catching your own.

↓ Prior to processing to supply shark fin soup markets, fins of Blue Sharks (*Prionace glauca*) sun-dry on racks outside a shark fin production factory in Kesennuma, Japan.

DO NOT TRY THIS AT HOME

Preparing shark fins for shark fin soup is a multi-step process, starting with a week or more of processing that involves cleaning and peeling the fin and digesting away the connecting tissue. What remains are the fin rays, called ceratotrichia, which are composed primarily of a protein similar to that of human hair. The ceratotrichia are boiled and take on the appearance of colorless cellophane noodles, at which point they are ready to be used. Ceratotrichia are said to be flavorless. Faux shark fin soup—that is, made without ceratotrichia—is being more widely used as a sustainable alternative, but it has not yet been widely accepted by shark fin soup consumers.

SHARKS AND US

Dogfish tales

An example of shark overfishing on the Atlantic coast of the United States in the twentieth century epitomizes unsustainable fishing in general. In addition, it exemplifies the nearly catastrophic consequences that can result from using common names of species instead of their scientific names. On the plus side, it also shows the benefits of effective management of overfished species.

Different life histories

As long as the accepted common names of both included the word dogfish, they were treated by fishery regulation in the US as if they were the same species. Whereas the life history characteristics of the Smooth Dogfish (Dusky Smoothhound; *Mustelus canis*) allowed it to be fished, those of the Spiny Dogfish (*Squalus acanthias*) were more conservative and meant that overfishing of the species could (and did) occur. The lesson from this confusion: use scientific names for species.

One of the sharkiest places on the planet in terms of greatest biomass (as opposed to the highest diversity of sharks) is the northwest Atlantic Ocean, and specifically the highly productive Grand Banks and adjacent waters. There resides what may be the most numerically abundant shark species, the Spiny Dogfish or Spurdog (*Squalus acanthias*; page 264).

WHAT'S IN A NAME?

Until about 2002, Spiny Dogfish were heavily fished, primarily using gillnets, along the Atlantic coast of the United States, along with another small shark, the Dusky Smoothhound (notably, also commonly called the Smooth Dogfish). Demand for these species in the United States was low, but they were prized in Europe for the fish and chips industry. Both species were sometimes caught on the same gear, and had the unfortunate coincidence of being classified for fishery statistics in a single category, "dogfish," since this is part of the common names of both species.

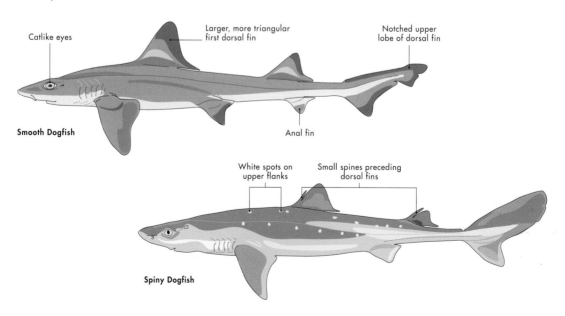

Categorizing these two species together, however, did not alter the hard fact that they have drastically different life history characteristics. Whereas Dusky Smoothhounds have fast life histories, maturing in three years and producing about 12–16 pups annually after a gestation period of about 10 months, Spiny Dogfish have slow life history characteristics. Spinies, as they are sometimes called, take as long as 20 years to mature and produce only 4–6 pups every other year, after a 24-month gestation period. Because of their very different life histories, Dusky Smoothhounds can support a fishery whereas Spiny Dogfish are less capable of doing so. Lumping the two species together therefore gave the mistaken impression that "dogfish" fisheries were thriving, when in fact the Spiny Dogfish stock was collapsing.

Luckily, fishery managers corrected the problem, imposing a temporary moratorium on fishing for Spiny Dogfish. The stock largely rebuilt and now supports a smaller fishery. Currently, the biggest United States shark fishery is for Dusky Smoothhounds, mainly off the coast of North Carolina. Most of the catch is exported to Australia and the United Kingdom as boneless fish fillets.

↑ A postage stamp from the UK featuring a Spiny Dogfish.

↑↑ A juvenile Smooth Dogfish, or Dusky Smoothhound, sits atop a ruler. This abundant species, which feeds on small fish, crustaceans, squid, bivalves, and worms, is commercially fished in the US at levels that do not constitute overfishing.

SHARKS AND US

Catch you later

Targeted fisheries for sharks constitute only a part of overfishing. The other fishery issue is bycatch, or untargeted or unwanted catch. Most fishing gear—including longlines, trawls, purse seines, and gillnets—catch sharks as bycatch. For example, high-seas drift gillnets in the North Pacific that target squid and salmon have large shark bycatch.

DEAD IN THE WATER

Mortality rates for shark bycatch can be substantial, even for sharks that are returned to the water alive. When caught on pelagic longlines, Smooth Hammerheads (*Sphyrna zygaena*), Scalloped Hammerheads (*S. lewini*), and Great Hammerheads (*S. mokarran*; page 44), as well as all three species of threshers, suffer mortality rates of 25–50 percent or more before being boated, and most released alive likely do not survive. Some species, on the other hand, do better on these same longlines, including Blue Sharks (*Prionace glauca*; page 138), Silky Sharks (*Carcharhinus falciformis*; page 272), and Oceanic Whitetip Sharks (*C. longimanus*; page 140), as well as makos. However, some of these same species suffer extremely high mortality in purse seines. Purse-seine bycatch in tropical tuna fisheries has been the primary driver in the decline of Silky Shark populations in both the Indian and Pacific Oceans. Even if released alive, deep-sea sharks likely experience extremely high mortality rates, regardless of the type of gear they are caught in.

↑ A moribund or dead Greenland Shark (*Somniosus microcephalus*) caught on a bottom trawl in the Northern Barents Sea, northeast Atlantic. Greenland Sharks are among the longest-lived vertebrates. In Iceland, they are consumed by humans after being fermented.

← It's not just a danger to sharks—among the catch on the deck of a trawler sits a sea turtle. In the US bottom shrimp trawl fishery, devices called Turtle Excluder Devices have been successful in diverting turtles, sharks, and larger fish from the net before they drown. Still, bottom trawls damage the seafloor and thus have impacts beyond bycatch.

WHO LIVES AND WHO DIES?

A variety of factors contribute to bycatch mortality, including water temperature, gear type, life stage of the shark, and the duration the shark has been hooked or entrapped. The physiology of each shark species underlies all of these factors. Although the physiology of bycatch mortality is poorly known, increases in metabolic end products (such as lactic acid) and stress hormones have been measured. On some of our experimental longlines, which are typically set for 45–60 minutes, mortality is lowest for Sandbar Sharks (*Carcharhinus plumbeus*; page 198), Bull Sharks (*C. leucas*; page 196), Lemon Sharks (*Negaprion brevirostris*; page 228), Tiger Sharks (*Galeocerdo cuvier*; page 232), and highest (especially in late summer) for Atlantic Sharpnose Sharks (*Rhizoprionodon terraenovae*; page 200) and Blacktip Sharks. Fortunately, the short soak times (the duration longlines are left in the water) minimize the number of sharks that die.

CONSERVATION LAW

Several international agreements protect sharks, including the Convention on International Trade in Endangered Species of Wild Fauna and Flora (CITES), a pact signed by 183 governments (including the European Union) that regulates international trade of threatened species, and the Convention on Migratory Species (CMS), an agreement among 131 member states to conserve migratory species throughout their ranges. The International Union for Conservation of Nature (IUCN) Shark Specialist Group prioritizes species at risk, monitors threats, and evaluates conservation action.

The primary law governing fisheries management in federal waters of the United States—that is, waters usually from 3 miles (5 km) of the coastline out to 200 miles (300 km)—is the Magnuson–Stevens Fishery Conservation and Management Act 1976. From the shore to a distance of 3 miles (5 km) or 9 miles (15 km) along the Gulf coast of Florida, management of marine resources comes under the jurisdiction of the states.

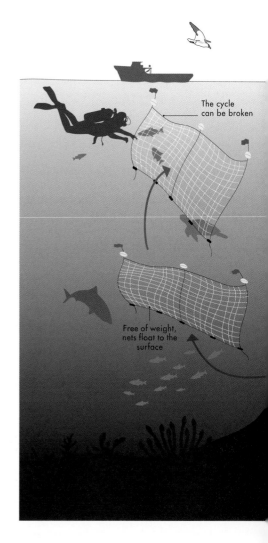

Ghost nets
A shocking amount of commercial fishing gear is lost or intentionally abandoned at sea. While all of it is potentially damaging, lost gear such as baited longline hooks and gillnets continue to catch animals, including sharks. Shown here are the latter, known as ghost nets. Floats on the top and weights on the bottom spread the nets open. The nets may foul corals and other sensitive organisms as they slowly disintegrate, a very slow and prolonged process that produces a whole new suite of environmental problems.

GHOSTLY TRAPS

Although accurate estimates are lacking, an unexpectedly large amount of abandoned or lost fishing gear fills the oceans. A study published in 2022 estimated that 62–219 million lb (28–100 million kg) of plastics were lost from industrial trawling, purse-seining, and pelagic longlines in 2018 alone. Known as ghost fishing gear, these hooks and nets continue to catch or trap and kill sharks and other endangered organisms, including marine mammals, seabirds, and sea turtles. Ghost fishing gear has been called the deadliest form of marine plastic.

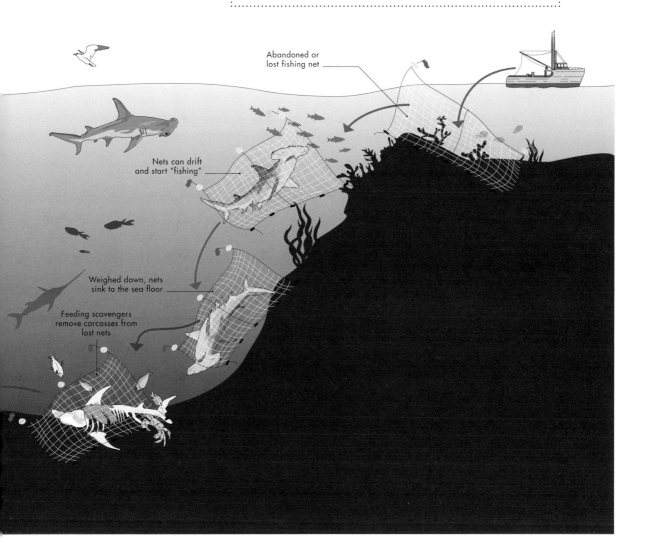

Climate change

Targeted fisheries and bycatch are considered the principal current dangers for sharks, but other major threats include climate change, habitat degradation and destruction, pollution (nutrient pollution from agricultural runoff and undertreated human sewage, plastic pollution, and other chemical pollution), exotic introductions, and aggregated human disturbance (persecution, noise pollution, and so on). Of these, the effects of climate change are least understood and have the widest potential consequences.

Human activity is causing global climate change—there is complete and unambiguous agreement about this among scientists who study the issue. There is, however, still some scientific uncertainty about many of the specifics, such as the time frame of impacts, the areas likely to be most impacted, and—most relevant to this discussion—how and when climate change will affect sharks, other living organisms, and Earth's marine and terrestrial ecosystems.

EFFECTS ON THE OCEANS

The aspects of climate change that very likely have already affected or will affect sharks (and other marine species) include elevated seawater temperatures, ocean acidification, changes in precipitation patterns that alter the salinity structure of nearshore and oceanic systems, increased intensity and frequency of tropical storms, rising sea levels that coastal wetland communities may not be able to keep pace with, and deoxygenation that may include increased size and severity of dead zones. Additionally, while some of these impacts will occur gradually, others may occur more suddenly.

← A Raja Ampat Epaulette Shark (*Hemiscyllium freycineti*) in Indonesia maneuvers on the bottom using its fins to crawl, a novel adaptation among sharks.

↗ The Gulf Stream, and feeder currents in the Gulf of Mexico and Caribbean. The Gulf Stream plays an important role in the transfer of warm and salty water from the tropics poleward. At the latitude of Greenland, prevailing westerly winds cool the current by extracting its heat, thus increasing its density and causing its water to sink. Since circulation in the North Atlantic depends on this rapid, intense cooling of salty water off Greenland, climate change impacts could be widespread and devastating as this large-scale circulation system is disrupted.

One looming impact of mind-boggling proportions is large-scale changes in ocean circulation, which could include a critical slowing of the Gulf Stream, an extremely powerful warm-water Atlantic Ocean current that affects the climate of the southeast United States as well as that of northwest Europe. Most studies of responses of sharks to climate change have examined the effects of elevated temperature, including spatial and temporal changes in migration, changes in the depth of the oxygen minimum layer, and impacts of these changes on metabolism, growth, reproduction, and foraging. Given the conservative life history characteristics of sharks as a group, including low fecundity and late maturity, adapting physiologically to the accelerated time frame of climate change impacts could be limited.

EFFECTS ON SHARK ECOLOGY

Climate change affects not only the physiology of sharks, but it also has potentially profound ecological ramifications. Many sharks are expected to migrate to higher latitudes or deeper waters as a result of temperature increases, since it is well known that temperature is a major driver of shark migrations.

A 2018 study, for example, found juvenile Bull Sharks in North Carolina estuaries that had not previously been frequented by the species and correlated their presence there with the early onset of summer temperatures. Ongoing population assessments for sharks in the United States are taking into account major shifts in suitable habitat for species such as the Scalloped Hammerhead and Blacktip Shark as migrations shift northward.

A move to higher latitudes may cause sharks to encounter ecosystems that are foreign or novel to them. This may lead to problems such as changes in the abundance and size structure of their populations, changes in their trophic structure, and changes in their behavior. It may also lead to mortalities and extirpations (regional extinctions), and even extinctions for species for which migrations may be difficult or improbable, or which may already be depleted or threatened by other stressors. Sharks inhabiting coastal ecosystems such as seagrass beds, mangroves, saltmarshes, and coral reefs may be disproportionately affected.

New developments and insights into climate change and other environmental impacts are occurring with breakneck rapidity, and studies on how these affect sharks and the ecosystems they depend on are also beginning to appear more regularly. However, identifying these problems is only the first step. Large-scale changes in how humans consume resources, especially fossil fuels, and how we preserve the biodiversity and integrity of ecosystems must follow, and quickly.

→ Scalloped Hammerheads (*Sphyrna lewini*) schooling near Cocos Island, Costa Rica, in the Pacific Ocean. These schools are part of larger, predominantly female, aggregations that occur in eastern tropical Pacific Ocean islands. An assessment of the species in 2018 concluded that it is critically endangered.

Coda: who is the real threat?

We began this book by acknowledging that some sharks, at some times and in some places, represent real threats to people. If you were an alien visiting the planet, you might conclude from the media coverage of sharks that prudence demanded you never dangle your toe in salt water, or even fresh water for that matter, lest some toothy shark devour you.

MARKETING MADNESS?

Because of the perception that sharks are real threats to bathers, surfers, snorkelers, and other water users, a profitable industry in personal shark repellents has arisen. Since, as we have already established, the threat of shark bites and attacks is vanishingly small, these devices are superfluous, and in any case they are of limited use at best or have no efficacy as shark deterrents. What's more, swimmers using shark deterrents mistakenly conclude that they do work since they don't get bitten or attacked, not realizing that they wouldn't have been bitten or attacked without them.

There are, however, three positives to personal shark deterrents. First, they provide a measure of confidence to bathers that allows them to enter the water. Second, the research behind the devices, at least in some cases, is rigorous, and that can lead to an increased understanding of shark sensory biology and behavior. Finally, some of the repellent technology may have meaningful applications, not so much to repel sharks from people (although methods that help shipwreck and plane crash victims avoid sharks would truly be beneficial), but to repel sharks from being caught as bycatch or to provide an alternative to beach nets that kill thousands of sharks annually.

The facts, however, tell a different story. The number of shark bites and attacks globally is low: 73 in 2021, according to the International Shark Attack Files. Despite the rarity of these events, many people perceive sharks as an existential threat, and to them an ocean without sharks might be acceptable. But that would be a short-sighted and foolish, even reckless, conclusion. Sharks swimming in the ocean may stoke our irrational fears and affect our behavior at the beach, but they are an essential part of the fabric of marine life that maintains a healthy ocean. Sharks play roles in marine, and some freshwater, ecosystems that we are only beginning to understand. Our actions, including overfishing, habitat alteration, changing the climate, etc., are much greater threats to the survival of many shark species in the short term, and to sharks as a group over a longer time frame, than sharks are to people.

Here ends our homage to sharks. In some small measure, we hope that we have provided you with knowledge about these magnificent predators you may not have known, or a newfound appreciation for them. If this book inspires you to act on their behalf, or on behalf of the ecosystems they inhabit, or to educate someone about the importance of thriving sharks to a healthy ocean and planet, then this effort will have been worthwhile. We live on a wondrous planet—one that needs its predators, including sharks. But these predators need you as well; clear-eyed, energetic, educated, and ready to appreciate, understand, and protect them.

↑ A non-sensationalized photo of a White Shark (*Carcharodon carcharias*) showing its large black eyes and white underside.

SQUALUS ACANTHIAS

Spiny Dogfish
Fish and chips shark

SCIENTIFIC NAME	*Squalus acanthias*
FAMILY	Squalidae
NOTABLE FEATURE	Slender shark, first dorsal fin small, spines on both dorsal fins, white spots on upper flanks
LENGTH	4 ft (1.2 m)
TROPHIC LEVEL	Opportunistic mesopredator of small fish and invertebrates

Despite being heavily overfished in the northwest Atlantic in the recent past and currently listed as Endangered in the Mediterranean Sea and throughout Europe, the Spiny Dogfish (also widely known as the Spurdog) remains plentiful in some other locations globally. Historically, the Spurdog has been considered one of the most abundant shark species. Its common name derives from the prominent spines, which are mildly toxic, preceding the two short dorsal fins. As noted in the previous chapter, the species gained notoriety in the United Kingdom and even globally when a 2021 survey caught specimens in the River Thames and the media sensationalized the finding by focusing on the species' venomous spines.

The Spiny Dogfish is found intertidally to depths of about 3,000 ft (900 m) and is one of only two shallow-water cold-temperate dogfish, as well as being the only Atlantic *Squalus* species that is also coastal. The sister species in the Pacific, the Pacific Spiny Dogfish (*S. suckleyi*), also has a coastal and deep-sea presence.

Reproduction in the Spiny Dogfish is yolk-sac viviparity (page 66), and it produces litters of usually 6-12 after a gestation period of 22–24 months, one of the longest among animals. The species is long-lived (40 years), with a late age of maturity at about 20 years. Because of historical overfishing, combined with the species' conservative life history characteristics, the IUCN lists the Spiny Dogfish as Vulnerable.

Spiny Dogfish are readily maintained in captivity and, owing to this, a disproportionate amount of the early knowledge of the biology of sharks as a group was based on findings from this species. This included information on the roles of fins, metabolic rates, thermal tolerance, digestion, age determination using the spines, and parasitic infections. However, there is one downside to this: Spiny Dogfish are generally representative only of species with similar phylogenetic histories, physiological systems, and ecological roles, and not all sharks.

In the 1930s, Spiny Dogfish were used as a source of vitamin A, a fat-soluble compound important to vision, the immune response, reproduction, and cellular communication. Fisheries for the Pacific Spiny Dogfish (*S. suckleyi*) off British Columbia, Canada, as well as for the School (or Tope) Shark (*Galeorhinus galeus*) off California, exploded during the Second World War when supplies of cod liver oil, the major source of vitamin A, were cut off. And if you have ever eaten fish and chips in the United Kingdom, you likely have consumed this species, which is known there as Rock Salmon.

→ Spiny Dogfish prowl the productive waters of the northwest Atlantic Ocean in search of potential prey. Distinguished by the dorsal fin spines and white spots on the flanks, the Spiny Dogfish is one of very few species of squalomorph sharks to live in coastal waters.

MUSTELUS CANIS

Dusky Smoothhound

Catlike eyes, pavement teeth

SCIENTIFIC NAME	*Mustelus canis*
FAMILY	Triakidae
NOTABLE FEATURE	Slender, elongated oval catlike eyes, flattened teeth, triangular dorsal fins, notched upper lobe of caudal fin
LENGTH	5 ft (1.5 m)
TROPHIC LEVEL	Mesopredator of crustaceans, worms, mollusks, squid, and small fish

Commonly called Smooth Dogfish to distinguish it from the Spiny Dogfish, the Dusky Smoothhound is not a dogfish at all. The species is common in the northwest Atlantic and has two subspecies.

The subspecies *Mustelus canis canis* is most common along the shallow continental shelf from New York to South Carolina, where it pups in shallow estuaries. It is also found in deeper waters, and was found to pup at 1,300 ft (400 m) in the Gulf of Mexico. This discovery means that this subspecies uses both shallow- and deep-water depth regimes for giving birth, a very unusual activity among sharks. *Mustelus canis insularis* is the island form of the species, and is found in deeper waters in the tropics.

Like Spiny Dogfish, Dusky Smoothhounds do well in captivity and have been used to uncover knowledge of shark biology, for which reason they have been called by some the lab rat of shark research. In particular, they have been used to study the role of olfaction (smell) in locating prey, heart function, hormonal control of skin pigmentation, responses to high levels of carbon dioxide associated with climate change and ocean acidification, and the roles of fins.

Also like the Spiny Dogfish, the Dusky Smoothhound is very abundant. Unlike that species, however, it has fast life history characteristics, including giving birth to about 12–16 young annually after about a 10-month gestation period. The age of maturity for males and females is, respectively, as early as two and three years. The species is thus capable of supporting a well-managed fishery. Even so, the Dusky Smoothhound is classed as Near Threatened on the IUCN Red List.

→ Dusky Smoothhounds use a combination of electroreception and olfaction to pinpoint buried prey, often small crabs, which are then crushed with their cobblestone-shaped teeth.

SQUATINA SQUATINA

Angel Shark

Ambush predator and ray doppelgänger

SCIENTIFIC NAME	*Squatina squatina*
FAMILY	Squatinidae
NOTABLE FEATURE	Flattened (depressed) body, terminal mouth, eyes on top of head, nasal barbels, large pectoral fins, prominent spiracles without valves
LENGTH	7.9 ft (2.4 m)
TROPHIC LEVEL	Ambush predator of bottom-dwelling bony fishes, sharks, and rays

If at first glance you identified the Angel Shark as a ray, you could be excused for the indiscretion. Although *Squatina squatina* and its 21 fellow members in the genus *Squatina* resemble batoids, with a flattened body and expanded pectoral fins (which are not connected to the head), and they have some ecological similarities to batoids, they are true sharks. As a group, angel sharks live coastally in temperate seas and in deeper waters in the tropics. *Squatina squatina*, whose common name is Angel Shark, is found in the northeast Atlantic and Mediterranean and Black Seas.

Angel Shark reproduction is by yolk-sac viviparity, and as many as 25 pups are born after a 10-month gestation period every other year. The 22 *Squatina* species are the only elasmobranchs with a hypocercal caudal fin—an inverse heterocercal tail, in which the lower lobe is larger than the upper. They are well-camouflaged lie-and-wait predators, which elevate their anterior with a stroke or two of their powerful tail as they ambush their prey and consume it using the large gape of their terminal mouth. The typical heterocercal fin characteristic of sharks, with a larger upper lobe, would be a disadvantage in this situation since it would tend to send the head in the wrong direction.

Angel Sharks have been fished commercially since the 1980s (they may be called Monkfish in some locations). Once very common, this species is now listed as Critically Endangered by the IUCN in the northeast Atlantic and Mediterranean and Black Seas, with declines due to historical and ongoing overfishing. The species has been extirpated from the North Sea and parts of the northern Mediterranean Sea. The only relatively healthy, protected since 2019, population occurs in the Canary Islands, and there are early signs of recovery of a population off the coast of Wales in the United Kingdom.

→ An Angel Shark amid the shallow seagrass community of the Canary Islands in the eastern Atlantic Ocean. Although the species is critically endangered, the population in the Canary Islands is not as depleted as in other areas, and it is protected there.

SHARKS AND US

CARCHARHINUS OBSCURUS

Dusky Shark
Poster child for depleted US sharks

SCIENTIFIC NAME	*Carcharhinus obscurus*
FAMILY	Carcharhinidae
NOTABLE FEATURE	Large body, interdorsal ridge, short snout, first dorsal fin begins slightly behind the rear of the pelvic fins
LENGTH	12.4 ft (3.8 m)
TROPHIC LEVEL	Generalist predator of bony fishes, sharks, rays, cephalopods, and other invertebrates

The Dusky Shark, found globally in warm-temperate and tropical waters of the continental shelf, is so closely related to the Galapagos Shark (*Carcharhinus galapagensis*) that some geneticists have considered them to be a single species. Their primary difference is their habitat. Whereas the Dusky Shark is closely associated with continents, the Galapagos Shark frequents islands. One additional, key, difference is that the Dusky Shark remains one of the most overfished sharks on the East Coast of the United States and is considered Endangered by the IUCN.

The genus *Carcharhinus* includes some larger sharks, such as the Bull Shark, Oceanic Whitetip Shark, and Bignose Shark (*C. altimus*). The biggest, however, is the Dusky Shark, exceeding 12 ft (3.6 m) in length. The biology of the Dusky Shark is not as well known as that of some of these other species, in part because of decreases in its abundance, but also because it is difficult to distinguish from other large carcharhinid sharks (note how vague our description is above). In general, assessing the populations of coastal sharks is fraught with problems, including estimating natural mortality, having accurate catch statistics, and knowing post-release mortality.

Like the Sandbar Shark, the Dusky Shark was prized for its fins, which has led to overfishing and consequent large population declines in both species. In 1993, the Fisheries Management Plan for Sharks of the Atlantic Ocean, which imposed restrictive quotas for these and other sharks, was implemented for the US gulf and east coasts by the US National Oceanic and Atmospheric Administration (NOAA). While the Sandbar Shark is on a trajectory to recover in the western North Atlantic by 2070, full recovery of the Dusky Shark is not expected until after 2100 despite their landings being prohibited since 1999. The reason it will take so long for these species to recover is due to their life history characteristics. Members of the "Large Coastal Shark" group in the fishery management plan, including the Lemon, Sandbar, and Dusky Sharks, have population doubling times of 20–25 years. In contrast, "Small Coastal Sharks," a category that includes the Dusky Smoothhound, Bonnethead (*Sphyrna tiburo*; page 204), and Atlantic Sharpnose Shark, all have population doubling times of 5–10 years.

The particularly slow recovery of Dusky Sharks is due to their three-year reproductive cycle. In contrast, Sandbar and Blacktip Sharks take only one year off after giving birth. At one time, Dusky Sharks were numerous off the coast of the southeastern United States. Their full recovery would be a major conservation success story.

→ Dusky Sharks, among the largest of the requiem sharks, are highly vulnerable to overfishing due to their conservative life history.

CARCHARHINUS FALCIFORMIS
Silky Shark
Smooth operator

SCIENTIFIC NAME	Carcharhinus falciformis
FAMILY	Carcharhinidae
NOTABLE FEATURE	Slender body, interdorsal ridge, second dorsal with elongated free tip, long sickle-shaped pectoral fins, very small pelvic fins
LENGTH	11.5 ft (3.5 m)
TROPHIC LEVEL	Generalist predator of small fish, squid, and pelagic crabs

Found worldwide in the tropics, the Silky Shark—so named because its small dermal denticles are smooth to the touch—is one of the most abundant surface-dwelling open-ocean sharks. It is also found in large schools around seamounts and at the edge of continental shelves in water less than 650 ft (200 m) deep. Silky Sharks trail only Blue Sharks as the most commonly caught shark species in commercial fisheries. The species is listed as Vulnerable on the IUCN Red List and has a decreasing population trend.

Silky Shark maturity varies by location and may range from five to 15 years. Litter size averages more than six pups and can reach 18, and the young are born after a 9–12-month gestation period. Again, depending on the region, females may have litters annually or every other year. The population doubling time has been estimated at between 4.5 and 14 years, a relatively fast time that accounts in part for the slightly rosier assessment of the species' conservation status compared to its congeners the Sandbar and Dusky Sharks.

In some areas, Silky Sharks are overfished along with Oceanic Whitetip Sharks as bycatch in purse-seine tuna fisheries. Both species, as well as Blue Sharks, also constitute bycatch in pelagic longline fisheries. Most of the sharks that end up in the international fin trade are bycatch in pelagic longline fisheries and there is concern that these species may become depleted. Interestingly, Silky Sharks released from pelagic longlines appear to have significantly higher survival rates than those released from tuna purse seines, with mortality in the latter estimated at 80 percent or more.

One solution to reducing shark mortality on longlines is changing the depth of the hooks. Silky and Oceanic Whitetip Sharks spend most of their time in the upper mixed layer, at depths shallower than 330 ft (100 m). Increasing the depth of hooks to that level or deeper has been shown to reduce bycatch rates significantly for these and most other shark species.

→ Two Silky sharks glide through the deep blue waters near Jardines de la Reina, Cuba.

GLOSSARY

↑ The setting sun sublimely silhouettes several serene Silky Sharks swimming slowly in the cerulean sea.

ampullae of Lorenzini Sensory receptors concentrated on the head of sharks and batoids capable of detecting weak electric fields of other organisms and even inanimate objects; used to guide the shark or batoid to its prey in the final stages of predation.

anal fin Median fin on the posterior ventral surface of many sharks and present as a fin fold in some batoids; absent in all members of the shark orders Squaliformes, Squatiniformes, Pristiophoriformes, and Echinorhiniformes. The specific function of the anal fin is not precisely known.

apex predator Animals, typically large, at the top of the food chain, with no or extremely few natural predators as adults.

aplacental viviparity Live birth without a placenta. An outdated term that typically refers to yolk-sac viviparity.

batoids Skates and rays; a group of more than 600 species of dorsoventrally (top-to-bottom) flattened fishes that share a common ancestor with sharks, with which they are classified as elasmobranchs.

benthic Of or relating to the bottom of the ocean or the seafloor.

biodiversity Refers broadly to the variety and number of species in an ecosystem.

biofluorescence Phenomenon in which organisms absorb higher-energy visible light wavelengths (e.g., blue) and emit light of lower energy levels (e.g., red, orange, or green). The Swell Shark (*Cephaloscyllium ventriosum*) and Chain Catshark (*Scyliorhinus retifer*) exhibit bright green biofluorescence.

bioluminescence Light emitted from living organisms using chemical reactions. The common name of the family Etmopteridae is lantern sharks, so named because the 51 species possess bioluminescent organs called photophores on the underside of the head.

bony fishes The dominant aquatic vertebrates, with approximately 35,000 kinds, including tunas, mackerels, mullets, cod, sturgeons, bass, etc. Bony fishes have a skeleton composed of bone, which is dense, rigid, hard tissue, and are in the class Osteichthyes (*oste* = bone, *ichthyes* = fish).

brackish Aquatic environments with a mixture of fresh and salt water. Brackish environments are diluted by rainfall, runoff, and/or input of fresh water from springs or rivers, and thus their salinity is lower than that of the open ocean or nearshore environment.

buccal pumping Mechanism of irrigating the gills with water drawn through the mouth after expanding the cheeks. It is widely practiced by sedentary fishes.

bycatch Untargeted or unwanted catch in a fishery, including organisms that are undersized, are of low commercial value, or are protected.

caudal fin Tailfin of a fish; the main propulsive fin of sharks.

cephalofoil The laterally expanded head of hammerhead sharks, so named because of the perception that it provides dynamic lift in the same manner as airfoils (wings) of airplanes.

chimaeras Also known as ghost sharks and holocephalans, chimaeras (order Chimaeriformes) are a group of more than 50 species of cartilaginous fishes classified along with sharks and batoids in the class Chondrichthyes.

Chondrichthyes Taxonomic class of fishes comprising the sharks, batoids, and chimaeras, all of which have skeletons composed of cartilage (*chondr* = cartilage).

claspers Rearward tubular extensions of the inner margin of the pelvic fins responsible for sperm transfer in sharks, batoids, and chimaeras; a distinguishing feature of this group.

cloaca Common opening for the urinary, genital, and intestinal tracts in chondrichthyan fishes and many other vertebrates.

continental shelf The seaward extension of the continent underlain by continental crust and covered by the ocean. Its width ranges from a few miles to hundreds of miles and marks the shoreline during the last ice age. Its seaward border is the steeper continental slope, which plunges to the ocean floor.

GLOSSARY

deep sea That part of the ocean deeper than 660 ft (200 m).

demersal Referring to organisms living on or near the seafloor (the demersal zone).

diel Having a period of 24 hours, day–night.

dimorphism Literally "two forms," dimorphism refers to differences in size, external features, shapes, or coloration of organisms of the same species. Among sharks, the most frequent reference is to sexual dimorphism.

dorsal fin Median fin(s) on the dorsal surface of fishes. In sharks, there may be one or two dorsal fins. Acts primarily as a keel, allowing quick turns and controlling roll.

dorsum Anatomical term referring to the upper surface or back of an animal.

ectotherm Animals whose body temperature is determined primarily by the temperature of the environment (*ekto* = out, *therm* = temperature). Most bony fishes and chondrichthyans are ectotherms.

embryophagy Phenomenon in which the fastest-growing embryo in each uterus eats its siblings, until only one remains in each uterus; a form of nutrition also called adelphophagy ("sibling eating") or intrauterine cannibalism. Found in the Sand Tigers (*Carcharias taurus*).

embryotrophy A novel form of embryonic nutrition occurring only in Tiger Sharks (*Galeocerdo cuvier*), in which embryos are sequestered in sacs containing a clear nutritive fluid called embryotroph that is secreted by the mother.

endotherm So-called "warm-blooded" animals, whose body temperature is elevated above that of the environment by trapping metabolic heat they generate using anatomical and physiological adaptations. Members of only five bony fish families and two shark families (Lamnidae, five species of mackerel sharks; and Alopiidae, three species of threshers) are endothermic.

epipelagic zone That part of the ocean shallower than 660 ft (200 m); also called the photic zone, since it marks the limit of sufficient light penetration for photosynthesis.

euryhaline Referring to organisms capable of tolerating a wide (*eury* = wide) variety of salinities. Euryhaline sharks and batoids include the Bull Shark (*Carcharhinus leucas*), Atlantic Stingray (*Hypanus sabinus*), and pristids (sawfish), as well as juveniles of some species, e.g. Sandbar Sharks (*C. plumbeus*).

fecundity The number of offspring per reproductive cycle or over an animal's lifetime.

fusiform Spindle-shaped, with a wider middle and tapering at both ends, as in members of the family Lamnidae (mackerel sharks).

gillnet Type of fishing gear consisting of a wall of nylon net suspended in the water column that ensnares fish and other organisms that swim into it; responsible for more shark (as well as turtle and marine mammal) bycatch mortality than any other fishing gear.

gill raker Non-respiratory processes projecting from the gill arches of fish. Used to trap small organisms in planktivorous sharks.

International Union for Conservation of Nature (IUCN) Global non-governmental authority on the status of the natural world and the measures needed to safeguard it. Founded in 1948, the IUCN is best known for its Red List of Threatened Species. The IUCN Shark Specialist Group (SSG) was established in 1991 to identify and prioritize shark species at risk, monitor threats, and evaluate conservation action.

longline Type of fishing gear consisting of a main line as long as 60 miles (100 km), with more than 1,000 baited hooks suspended from branch lines.

mesopelagic zone That part of the deep sea at depths of 540–3,300 ft (165–1,000 m); also called the twilight zone, since light penetration at the top of the zone is only about 1 percent of that at the surface. The mesopelagic zone is one of the planet's largest living spaces.

mesopredator Organisms between the extremes of the food web; that is, intermediate between herbivores (the primary consumers) and apex predators (*meso* = middle). Mesopredators are typically smaller than apex predators and can be on their menus. Coral reef mesopredators include larger sharks such as Caribbean Reef Sharks (*Carcharhinus perezi*) and smaller sharks such as Epaulette Sharks (*Hemiscyllium ocellatum*).

neonate Newborn organism.

neritic zone Relatively shallow zone above the continental shelf from the low-tide mark to the shelf's edge.

oophagy Egg-eating, referring to the phenomenon of embryos of lamniform sharks and false catsharks, once they have used the stored energy of their yolk sac, swimming within the uteri and consuming ovulated ova.

oviparity Egg-laying, in which the source of nutrition throughout the entire developmental period is the yolk deposited by the mother. About 40 percent of sharks and batoids are oviparous.

parthenogenesis Virgin birth; that is, development of an egg into an embryo without fertilization. Occurs in many groups (e.g., some plants, insects, bony fishes, and other non-mammalian vertebrates) as a natural form of asexual reproduction. In sharks, parthenogenesis was first reported in a captive Bonnethead (*Sphyrna tiburo*) and has now been described in several galeomorph shark species and one eagle ray.

pectoral fins Paired lateral fins of fishes, positioned toward the anterior. Thought to control pitch, roll, and possibly yaw, and may provide dynamic lift. The equivalents in other vertebrate groups are arms and wings.

pelagic Zone of ocean (or organisms inhabiting this zone) above the bottom, as opposed to benthic or demersal; also called the water column.

pelvic fins Paired, lateral fins of fishes, positioned toward the posterior. They primarily control pitch.

photic zone *See* epipelagic zone.

photophore Bioluminescent (light-producing) organ present in many organisms, including some sharks and other fishes, and cephalopods and many other groups of invertebrates.

phylogeny The evolutionary history of an organism or group.

planktivore Plankton-eating organism.

plankton Diverse group of aquatic organisms, including plants (phytoplankton) and animals (zooplankton), typically small, that cannot propel themselves against a current; an important food source for planktivorous organisms.

prey release Phenomenon in which removal or reduction in the population of a predator in an ecosystem results in increases in the population of its prey and subsequent decreases in the prey's prey. Also called predator release.

purse seine Type of fishing gear consisting of a seine net with a float line at the top and a lead line at the bottom, positioned to encircle a school of fish (such as tuna), with one or two boats pulling it from the larger harvest vessel. Once the float line circle is closed, the opening at the bottom of the net is closed, or pursed, by cinching the lead line and thus preventing fish from escaping. Bycatch in purse-seine fisheries is mainly Silky Sharks (*Carcharhinus falciformis*), Oceanic Whitetips (*C. longimanus*), and mantas, with other species caught episodically.

ram ventilation Mode of respiration in which the fish swims constantly with its mouth ajar so that sufficient water passes over their gills to oxygenate them.

rectal gland Finger-like or bean-shaped structure that eliminates the excess sodium and chloride that diffuse into a shark's body as a result of the higher concentrations of these two ions in seawater compared to the internal fluid environment of the shark; a distinguishing feature of sharks, batoids, and chimaeras, as well as some bony fishes.

rostrum Snout of sharks or batoids, typically underlain with soft, even spongy, cartilage that absorbs some of the forces associated with movement of the jaws during feeding.

salinity Saltiness of water. Oceanic water has a salinity of about 35, whereas rivers and lakes have salinities near zero or in the low single digits.

GLOSSARY

speciation The phenomenon of forming a new species. One mechanism by which species form involves conditions that isolate members of the same species from one another for long periods of time under different environmental conditions.

spiracles Paired bilateral openings behind the eyes that connect the mouth to the water environment in batoids and some sharks. In the requiem sharks (family Carcharhinidae), the spiracles are vestigial, meaning that the structure is rudimentary and has lost its core function.

swim bladder Gas-filled internal structure found in most bony fishes, but absent in sharks, that allows them to adjust their buoyancy and in the process save energy and expand their range.

thermocline Transition zone between the warmer upper layer and the cold deep water below. At about 3,300 ft (1,000 m) in most of the world's oceans, the temperature has generally declined to 36–41°F (2–5°C).

trophic cascade Phenomenon resulting from predator (or prey) release across three or more trophic levels.

twilight zone *See* mesopelagic zone.

ventrum Anatomical term referring to the lower surface or underside (belly) of an animal.

vertebrate Member of the phylum Chordata, subphylum Vertebrata (Craniata). The most prominent vertebrate characteristic is a skull and vertebral column (backbone). Vertebrates (craniates) also exhibit bilateral symmetry, a brain, and sensory organs.

viviparity Live birth, where the developing embryos are protected within the mother while they develop and grow, and are typically born at a size that increases their potential to survive.

yolk-sac viviparity Form of live birth where the developing pup receives the bulk of its nourishment from its yolk sac with little or no nourishment from the mother. Overall, about 40 percent of extant species of elasmobranchs practice yolk-sac viviparity.

RESOURCES

BOOKS

Abel, D. C. and R. D. Grubbs. *Shark Biology and Conservation: Essentials for Educators, Students, and Enthusiasts* (Johns Hopkins University Press, 2020)

Carrier, J. C., J. A. Musick, and M. R. Heithaus, eds. *Sharks and Their Relatives II: Biodiversity, Adaptive Physiology, and Conservation* (CRC Press, 2010)

Carrier, J. C., J. A. Musick, and M. R. Heithaus, eds. *Biology of Sharks and Their Relatives* (CRC Press, 2012)

Carrier, J. C., C. A. Simpfendorfer, M. R. Heithaus, and K. E. Yopak, eds. *Biology of Sharks and Their Relatives*. 3rd edition. (CRC Press, 2022)

Castro, J. I. *The Sharks of North America* (Oxford University Press, 2010)

Ebert, D. A., M. Dando, and S. Fowler. *Sharks of the World: A Complete Guide* (Princeton University Press, 2021)

Hamlett, W. C., ed. *Sharks, Skates, and Rays: The Biology of Elasmobranch Fishes* (Johns Hopkins University Press, 1999)

Helfman, G. and G. Burgess. *Sharks: The Animal Answer Guide* (Johns Hopkins University Press, 2014)

Klimley, A. P. *The Biology of Sharks and Rays* (University of Chicago Press, 2013)

Last, P. R. and J. D. Stevens. *Sharks and Rays of Australia* (CSIRO, 2009)

Last, P., G. Naylor, B. Séret, W. White, M. de Carvalho, and M. Stehmann, eds. *Rays of the World* (CSIRO Publishing, 2016)

Musick, J. A. and B. McMillan. *The Shark Chronicles: A Scientist Tracks the Consummate Predator* (Macmillan, 2003)

Peschak, T. P. *Sharks and People: Exploring Our Relationship with the Most Feared Fish in the Sea* (University of Chicago Press, 2014)

Shiffman, D. *Why Sharks Matter: A Deep Dive with the World's Most Misunderstood Predator* (Johns Hopkins University Press, 2022)

SCIENTIFIC JOURNAL ARTICLES

Abel, D. C., R. F. Young, J. A. Garwood, M. Travaline, and B. K. Yednock. "Observations on the importance of salinity structure to shark populations and nurseries in two South Carolina estuaries." *American Fisheries Society Symposium* 50: 109–124 (2007)

Abrantes, K. G., J. M. Brunnschweiler, and A. Barnett. "You are what you eat: Examining the effects of provisioning tourism on shark diets." *Biological Conservation* 224: 300–308 (2018)

Arnt, J. L. "Residency, diel movement, and tidal patterns of large juvenile Bull Sharks (*Carcharhinus leucas*) in Winyah Bay, SC." M.Sc. thesis, Coastal Carolina University (2020)

Bangley, C. W., L. Paramore, D. S. Shiffman, and R. A. Rulifson. "Increased abundance and nursery habitat use of the Bull Shark (*Carcharhinus leucas*) in response to a changing environment in a warm-temperate estuary." *Scientific Reports* 8: 1–10 (2018)

Barley, S. C., M. G. Meekan, and J. J. Meeuwig. "Diet and condition of mesopredators on coral reefs in relation to shark abundance." *PLoS ONE* 12 (4): e0165113 (2017). https://doi.org/10.1371/journal.pone.0165113

Bond, M. E., J. Valentin-Albanese, E. A. Babcock, N. E. Hussey, M. R. Heithaus, and D. D. Chapman. "The trophic ecology of Caribbean Reef Sharks (*Carcharhinus perezi*) relative to other large teleost predators on an isolated coral atoll." *Marine Biology* 165 (4): 1–13 (2018)

Bouyoucos, I. A., M. Romain, L. Azoulai, K. Eustache, J. Mourier, J. L. Rummer, and S. Planes. "Home range of newborn Blacktip Reef Sharks (*Carcharhinus melanopterus*), as estimated using mark-recapture and acoustic telemetry." *Coral Reefs* 39 (5): 1209–1214 (2020)

Chapman, D. D., E. A. Babcock, S. H. Gruber, J. D. Dibattista, B. R. Franks, S. A. Kessel, T. Guttridge et al. "Long-term natal site-fidelity by immature Lemon Sharks (*Negaprion brevirostris*) at a subtropical island." *Molecular Ecology* 18 (16): 3500–3507 (2009)

Clark, E. "Instrumental conditioning of Lemon Sharks." *Science*, 130 (3369): 217–218 (1959)

Collatos, C., D. C. Abel, and K. L. Martin. "Seasonal occurrence, relative abundance, and migratory movements of juvenile Sandbar Sharks, *Carcharhinus plumbeus*, in Winyah Bay, South Carolina." *Environmental Biology of Fishes* 103: 859–873 (2020)

Comfort, C. M. and K. C. Weng. "Vertical habitat and behaviour of the Bluntnose Sixgill Shark in Hawaii." *Deep Sea Research Part II: Topical Studies in Oceanography* 115: 116–126 (2015)

Cortés, E. "Standardized diet compositions and trophic levels of sharks." *ICES Journal of Marine Science* 56 (5): 707–717 (1999)

Cortés, E. and S. H. Gruber. 1990. "Diet, feeding habits and estimates of daily ration of young Lemon Sharks, *Negaprion brevirostris* (Poey)." *Copeia* 1990 (1): 204–218 (1990)

Cotronei, S., K. Pozo, O. Audy, P. Přibylová, and S. Corsolini. "Contamination profile of DDTs in the shark *Somniosus microcephalus* from Greenland seawaters." *Bulletin of Environmental Contamination and Toxicology* 101 (1): 7–13 (2018)

De Necker, L. "The trophic dynamics of the Broadnose Sevengill Shark (*Notorynchus cepedianus*) in False Bay, South Africa, using multiple tissue stable isotope analysis." M.Sc. thesis, University of Cape Town (2017)

Dhellemmes, F., J. S. Finger, M. J. Smukall, S. H. Gruber, T. L. Guttridge, K. L. Laskowski, and J. Krause. "Personality-driven life history trade-offs differ in two subpopulations of free-ranging predators." *Journal of Animal Ecology* 90 (1): 260–272 (2021)

Edrén, S. M. C. and S. H. Gruber. "Homing ability of young Lemon Sharks, *Negaprion brevirostris*." *Environmental Biology of Fishes* 72 (3): 267–281 (2005)

Frisch, A. J., M. Ireland, J. R. Rizzari, O. M. Lönnstedt, K. A. Magnenat, C. E. Mirbach, and J. P. A. Hobbs. "Reassessing the trophic role of reef sharks as apex predators on coral reefs." *Coral Reefs* 35 (2): 459–472 (2016)

Gelsleichter, J. and N. J. Szabo. "Uptake of human pharmaceuticals in Bull Sharks (*Carcharhinus leucas*) inhabiting a wastewater-impacted river." *Science of the Total Environment* 456: 196–201 (2013)

Grubbs, R. D., J. A. Musick, C. L. Conrath, and J. G. Romine. "Long-term movements, migration, and temporal delineation of a summer nursery for juvenile Sandbar Sharks in the Chesapeake Bay region." *American Fisheries Society Symposium* 50: 87–107 (2007)

Guttridge, T. L. and C. Brown. "Learning and memory in the Port Jackson Shark, *Heterodontus portusjacksoni*." *Animal Cognition* 17 (2): 415–425 (2014)

Guttridge, T. L., S. H. Gruber, K. S. Gledhill, D. P. Croft, D. W. Sims, and J. Krause. "Social preferences of juvenile Lemon Sharks, *Negaprion brevirostris*." *Animal Behaviour* 78 (2): 543–548 (2009)

Hammerschlag, N., L. Williams, M. Fallows, and C. Fallows. "Disappearance of White Sharks leads to the novel emergence of an allopatric apex predator, the Sevengill Shark." *Scientific Reports* 9 (1): 1–6 (2019)

Heithaus, M., L. Dill, G. Marshall, and B. Buhleier. "Habitat use and foraging behavior of Tiger Sharks (*Galeocerdo cuvier*) in a seagrass ecosystem." *Marine Biology* 140 (2): 237–248 (2002)

Keller, B. A., J. S. Finger, S. H. Gruber, D. C. Abel, and T. L. Guttridge. "The effects of familiarity on the social interactions of juvenile Lemon Sharks, *Negaprion brevirostris*." *Journal of Experimental Marine Biology and Ecology* 489: 24–31 (2017)

Keller, B. A., N. F. Putman, R. D. Grubbs, D. . Portnoy, and T. P. Murphy. "Map like use of Earth's magnetic field in sharks." *Current Biology* 31 (13): 2881–2886 (2021)

Kuczenski, B., C. Vargas Poulsen, E. L. Gilman, M. Musyl, R. Geyer, and J. Wilson. "Plastic gear loss estimates from remote observation of industrial fishing activity." *Fish and Fisheries* 23 (1): 22–33 (2022)

Laxson, C. J., N. E. Condon, J. C. Drazen, and P. H. Yancey. "Decreasing urea: trimethylamine N-oxide ratios with depth in chondrichthyes: a physiological depth limit?" *Physiological and Biochemical Zoology* 84 (5): 494–505 (2011)

Leigh, S. C., Y. P. Papastamatiou, and D. P. German. "Seagrass digestion by a notorious 'carnivore'." *Proceedings of the Royal Society B: Biological Sciences* 285 (1886), 20181583 (2018). doi: 10.1098/rspb.2018.1583

Leigh, S. C., Y. P. Papastamatiou, and D. P. German. "Gut microbial diversity and digestive function of an omnivorous shark." *Marine Biology* 168 (5): 1–16 (2021)

MacNeil, M. A., B. C. McMeans, N. E. Hussey, P. Vecsei, J. Svavarsson, K. M. Kovacs, C. Lydersen et al. "Biology of the Greenland Shark *Somniosus microcephalus*." *Journal of Fish Biology* 80 (5): 991–1018 (2012)

Maljković, A. and I. M. Côté. "Effects of tourism-related provisioning on the trophic signatures and movement patterns of an apex predator, the Caribbean Reef Shark." *Biological Conservation* 144 (2): 859–865 (2011)

Natanson, L. J. and G. B. Skomal. "Age and growth of the White Shark, *Carcharodon carcharias*, in the western North Atlantic Ocean." *Marine and Freshwater Research* 66 (5): 387–398 (2015)

Nelson, D. R. "On the field study of shark behavior." *American Zoologist* 17 (2): 501–507 (1977)

Niella, Y., V. Raoult, T. Gaston, K. Goodman, R. Harcourt, V. Peddemors, and A. F. Smoothey. "Reliance of young sharks on threatened estuarine habitats for nutrition implies susceptibility to climate change." *Estuarine, Coastal and Shelf Science* 268: 107790 (2022)

Nielsen, J., R. B. Hedeholm, J. Heinemeier, P. G. Bushnell, J. S. Christiansen, J. Olsen, C. B. Ramsey et al. "Eye lens radiocarbon reveals centuries of longevity in the Greenland Shark (*Somniosus microcephalus*)." *Science* 353 (6300): 702–704 (2016)

Pullen, E. "Microplastics in the digestive system of the Atlantic Sharpnose Shark (*Rhizoprionodon terraenovae*) in Winyah Bay, SC." M.Sc. thesis, Coastal Carolina University (2019)

Ruppert, J. L., M. J. Travers, L. L. Smith, M.-J. Fortin, and M. G. Meekan. "Caught in the middle: Combined impacts of shark removal and coral loss on the fish communities of coral reefs." *PLoS ONE* 8: e74648 (2013). https://doi.org/10.1371/journal.pone.0074648

Sandin, S. A., B. J. French, and B. J. Zgliczynski. "Emerging insights on effects of sharks and other top predators on coral reefs." *Emerging Topics in Life Sciences* 6 (1): 57–65 (2022)

Shabetai, R., D. C. Abel, J. B. Graham, V. Bhargava, R. S. Keyes, and K. Witztum. "Function of the pericardium and pericardioperitoneal canal in elasmobranch fishes." *American Journal of Physiology* 1985: H198–209 (1985)

Shipley, O. N., J. W. Brownscombe, A. J. Danylchuk, S. J. Cooke, O. R. O'Shea, and E. J. Brooks. "Fine-scale movement and activity patterns of Caribbean Reef Sharks (*Carcharhinus perezi*) in The Bahamas." *Environmental Biology of Fishes* 101 (7): 1097–1104 (2018)

Speed, C. W., M. G. Meekan, D. Rowat, S. J. Pierce, A. D. Marshall, and C. J. Bradshaw. "Scarring patterns and relative mortality rates of Indian Ocean Whale Sharks." *Journal of Fish Biology* 72 (6): 1488–1503 (2008)

Tanaka, S., Y. Shiobara, S. Hioki, H. Abe, G. Nishi, K. Yano, and K. Suzuki. "The reproductive biology of the Frilled Shark, *Chlamydoselachus anguineus*, from Suruga Bay, Japan." *Japanese Journal of Ichthyology* 37 (3): 273–291 (1990)

Wingar, J. "Osmoregulation and salinity preference in juvenile Sandbar Sharks (*Carcharhinus plumbeus*) in Winyah Bay, SC, USA." M.Sc. thesis, Coastal Carolina University (2019)

USEFUL WEBSITES

Afrioceans Conservation Alliance
https://www.aoca.org.za

American Elasmobranch Society
https://elasmo.org

Chondrichthyan Tree of Life
https://sharksrays.org

European Elasmobranch Association
http://eulasmo.org

Florida Museum of Natural History Learn About Sharks
https://www.floridamuseum.ufl.edu/discover-fish/sharks

International Shark Attack File
https://www.floridamuseum.ufl.edu/shark-attacks

International Union for Conservation of Nature Red List of Threatened Species—Sharks
https://www.iucnredlist.org/search?query=sharks&searchType=species

Save Our Seas Foundation
https://saveourseas.com

Sharks4Kids
https://www.sharks4kids.com

Shark Trust
https://www.sharktrust.org/pages/category/discover-sharks

Sharkspeak
https://www.sophiemaycocksharkspeak.com

→ A single Lemon Shark (*Negaprion brevirostris*) breaks up the abrupt transition between sea and sand in shallow tropical waters, in the process creating a silhouette on the bottom.

RESOURCES

INDEX

A

acanthodians 34
acoustic telemetry 98–9
age 102
    Greenland Sharks 202
    sexual maturity 102, 104–5
alimentary migrations 123
Alopias superciliosus see Bigeye Thresher
Alopias vulpinus see Common Thresher
ampullae of Lorenzini 26, 52, 60, 82, 112, 172, 181
anal fins 50
Angel Shark 268–9
angel sharks 66–7, 244
anoxia 80
Apalachicola Bay 190
apex predators 13, 23, 86, 89–94, 105, 122, 190, 214–15
aplacental viviparity 144
Atlantic Sharpnose Shark 91, 98, 184, 190, 200–1, 214, 256
Azoic hypothesis 152

B

bamboo sharks 29, 50, 72, 210, 238–9
Basking Shark 23, 46–7, 90, 120, 122, 123, 129, 226, 249
batoids 14, 24, 28, 36, 112, 211
    see also rays; skates
Benchley, Peter 176
Bigeye Thresher 121, 130, 134, 142
Bimini Biological Field Station 215, 218–21
biofluorescence 42, 157
bioluminescence 42, 54, 78, 157, 164
biomedical applications 21
birth 66–7, 82, 105, 130
    see also yolk-sac viviparity

Blackmouth Catshark 172–3
Blacknose Shark 93, 184, 190, 214
Blacktip Reef Shark 22, 210, 214
Blacktip Shark 29, 60, 122, 184, 190, 214, 250, 256, 260
Blue Shark 65, 68, 101, 122, 129, 130, 132, 134, 138–9, 255
Bluntnose Sixgill Shark 32, 38–9, 52, 68, 86, 89, 101, 119
body types 125–9
Bonnetheads 104, 105, 184, 190, 204–5, 218
Broadnose Sevengill Shark 89, 225
buccal pumping 230
Bull Shark 23, 70, 86, 93, 178, 180–1, 184, 190, 195, 196–7, 214, 256, 260
buoyancy 26, 31, 51, 181, 228
bycatch 78, 96, 108, 112, 132–4, 136, 140, 162, 172, 194, 200, 204, 254–7, 272

C

capture stress 94
Carboniferous Period 34, 168
Carcharhinus acronotus see Blacknose Shark
Carcharhinus albimarginatus 86, 210
Carcharhinus amblyrhynchos 70
Carcharhinus brevipinna 89
Carcharhinus falciformis see Silky Shark
Carcharhinus galapagensis see Galapagos Shark
Carcharhinus isodon 190
Carcharhinus leucas see Bull Shark
Carcharhinus longimanus see Oceanic Whitetip Shark
Carcharhinus melanopterus see Blacktip Reef Shark

Carcharhinus obscurus see Dusky Shark
Carcharhinus perezi see Caribbean Reef Shark
Carcharhinus plumbeus see Sandbar Shark
Carcharias taurus see Sand Tiger
Carcharodon carcharias 22, 106–7
Caribbean Reef Shark 8–11, 72, 89, 93, 214, 234–5
carpet sharks 80, 238
cartilage 24, 32
catfish 28
cat sharks 42, 66, 68, 105, 154, 157, 160, 172–3, 211, 224
caudal (tail) fin 24, 50, 56, 268
Centrophorus uyato 170–1
Centroscymnus coelolepis see Portuguese Dogfish
cephalofoils 44
Cephaloscyllium ventriosum see Swell Shark
ceratotrichia 26, 251
Cetorhinus maximus see Basking Shark
Chesapeake Bay 98, 101, 104, 184, 187, 189
Chiloscyllium griseum 72
Chiloscyllium plagiosum see Whitespotted Bamboo Shark
chimaeras 14, 24, 28
Chlamydoselachus anguineus see Frilled Shark
Chondrichthyes 14, 24, 28
CITES 256
claspers 51, 65, 114
classification 14–16
climate change 194–5, 248, 258–61
Common Thresher 38, 120, 121, 142–3
conservation law 256
continental shelf 180, 189, 208–9

Convention on Migratory Species (CMS) 256
Cookiecutter Shark 119, 121, 129, 154, 157, 160, 164–5
coral reefs 31, 86–9, 93, 114, 209, 210–15, 222, 260
Coral Triangle 208
countercurrent exchange 57
counterillumination 54, 78
countershading 54
cow sharks 32, 50, 66–7, 160
crawling 52
Crocodile Shark 118, 129, 130
Cuban Dogfish 94

D

defensive behavior 70–1
dehydration 181
demersal zone 172
development 29, 66–7
DNA 32
dogfish sharks 23, 67, 105, 154, 160, 161, 252–3
dorsal fins 38, 50, 96, 102, 161, 198
Dusky Shark 215, 244–5, 270–1
Dusky Smoothhound 190, 250, 252–3, 266–7

E

ears 52, 59
Echinorhinus cookei 68
ecosystems 23, 86–93, 105, 243, 260
ecotourism 21, 234, 244
ectotherms 56
eggs 66–8, 172
embryophagy 236
embryotrophy 232
endotherms 56–7, 74, 110, 125
energy budget 218
environmental DNA (eDNA) 95, 105
Epaulette Shark 50, 52, 93, 214

INDEX

epaulette sharks 80
epipelagic zone 118, 120, 121, 122–3, 140, 153
Etmopterus spinax 54, 78–9
euryhaline species 184
evolution 18–47
eyelids 26, 60
eyes 38, 52, 60, 157

F

Falcatus falcatus 34
fecundity 28–9, 38, 68, 102, 105, 259
feeding *see* plankton-eating sharks; prey/predation
Finetooth Shark 190
fins 50–2
  anal 50
  caudal 24, 50, 56, 268
  claspers 51, 65
  dorsal 38, 50, 96, 102, 161, 198
  finning 250
  pelvic 50, 51, 52, 65
  shark fin soup 198, 250–1
fishing threat 122, 132–6, 162, 194, 244, 245, 250–7, 264, 266, 268, 270, 272
  Atlantic Sharpnose Shark 200
  Blackmouth Catshark 172
  Bonnethead 204
  gillnets 132, 194, 252, 254
  longline fishing 132, 134, 198, 255, 256, 257, 272
  lost gear 257
  Oceanic Whitetip Shark 140
  Porbeagle 110
  purse seines 255, 257, 272
  Sandbar Shark 198
  thresher sharks 142
  see also bycatch
fossil record 32–4
freshwater sharks 98, 174–205, 263

Frilled Shark 105, 152, 168–9
fusiform sharks 129

G

Galapagos Shark 210
Galeocerdo cuvier see Tiger Shark
galeomorphs 36
Galeorhinus galeus 96
Galeus melastomus 172–3
gestation periods 105, 114, 168, 232
ghost fishing gear 257
ghost sharks *see* chimaeras
gill arches 37, 38, 52
gill rakers 46, 108
gill slits 24, 38, 50, 56, 112
gillnets 132, 194, 252, 254
Ginglymostoma cirratum see Nurse Shark
Glyphis 180, 194
Goblin Shark 152, 166–7
Great Dying *see* Permian–Triassic extinction event
Great Hammerhead 44–5, 68, 86, 93, 94, 214, 215, 255
Great White Shark *see* White Shark
Greenland Shark 86, 102, 187, 202–3, 226, 228–9
Grey Bamboo Shark 72
Grey Reef Shark 70, 89, 93, 210, 214
growth 102, 138, 163
Gruber, Samuel "Sonny" 215, 221
gulper sharks 68, 105, 154, 163, 170–1
Gummy Shark 250

H

Halmahera Epaulette Shark 80–1
Haploblepharus edwardsii 224
head 52
hearing 59

heart 51, 60–1
Hemingway, Ernest 74
Hemiscyllium halmahera 80–1
Hemiscyllium ocellatum see Epaulette Shark
Heterodontus francisci 60
Heterodontus portusjacksoni see Port Jackson Shark
Hexanchus 154, 161
Hexanchus griseus see Bluntnose Sixgill Shark
histotroph 67
horn sharks 60, 66, 76
houndsharks 67, 211
human impacts 192–5, 240–72
  see also climate change; fishing threat; pollution
hybodonts 34
hypoxia 80, 192

I

Icelandic Catshark 157
individual recognition 70
internal fertilization 26, 51, 65
International Shark Attack File 21
International Union for Conservation of Nature (IUCN) 256
  Critically Endangered species 110, 140, 236, 244, 268
  Endangered species 144, 170, 198, 234, 244, 270
  Least Concern species 82, 136, 172
  Near Threatened species 78, 138, 232, 266
  Vulnerable species 142, 196, 230, 264, 272
intertidal zones 194–5, 209, 223
Isistius brasiliensis see Cookiecutter Shark

Isurus oxyrinchus see Shortfin Mako
Isurus paucus see Longfin Mako

J

jawless fishes 36
jaws 27, 31, 34, 35, 36–7
  bite force 76
  Bluntnose Sixgill Shark 38
  Goblin Shark 166
  Nurse Shark 40
  Tiger Shark 232

K

kitefin sharks 157, 160

L

Lamna ditropis see Salmon Shark
Lamna nasus see Porbeagle
lamniform sharks 108, 166
lantern sharks 78, 154, 157
lateral line 52, 60
learning 72
lecithotrophy 66
Lemon Shark 52, 70–2, 73, 101, 178, 184, 190, 195, 214, 215, 218–21, 222, 230–1, 256
Leopard Shark 50
Little Gulper Shark 170–1
liver 26, 51, 86, 106, 228
  oil 21, 46, 144, 163, 250, 264
living fossils 32–4, 37
Longfin Mako 121, 125
longline fishing 132, 134, 198, 255, 256, 257, 272
Longnose Sawshark 82–3

M

mackerel sharks 56, 67, 106, 125
Magnuson–Stevens Fishery Conservation and Management Act 256

285

mangroves 208, 209, 210, 222, 260
Matawan Creek 176, 178, 195
matrotrophy 66
Megachasma pelagios see Megamouth
Megalodon 37, 106
Megamouth 90, 108–9, 123, 129
mesopelagic zone 150, 153, 154
mesopredators 23, 89, 91, 93, 105, 190, 210, 214–15
migration 29, 98, 100–1, 122–3, 187
  alimentary 123
  Basking Shark 46
  climate change 259–60
  Lemon Shark 218, 220–1
  "roadkill" 249
  Salmon Shark 136
  White Shark 106, 122–3
  see also vertical migration
Mitsukurina owstoni see Goblin Shark
Mollisquama 157
Murrells Inlet 194
Mustelus antarcticus 250
Mustelus canis see Dusky Smoothhound
Myrberg, Arthur 140

N

Nebrius ferrugineus see Tawny Nurse Shark
Negaprion acutidens 214
Negaprion brevirostris see Lemon Shark
Nelson, Don 70
neritic zone 120
nictitating membranes 26, 60
nocturnal species 40
Notorynchus cepedianus see Broadnose Sevengill Shark
Nurse Shark 23, 40–1, 214

nurseries 101, 176, 184, 189, 218, 220

O

ocean circulation 259
Oceanic Whitetip Shark 23, 86, 118, 129, 130, 134, 140–1, 255
oophagy 67, 74, 106
Orcas 86, 89, 106, 225
Otodus megalodon 37

P

Pacific Sleeper Shark 226
parthenogenesis 204
passive integrated transponders (PIT) tags 98, 101
pectoral fins 35, 37, 50, 52
pelvic fins 50, 51, 52, 65
pericardium 61
Permian–Triassic extinction event 34
personality 73, 221
photophores 157
pitching 50
placental shark species 67, 68, 138
plankton-eating sharks 23, 46, 90, 105, 108, 123, 129, 144, 249
plastic pollution 192, 200, 248, 257
Pliotrema warreni 54
pocket sharks 157
polar coasts 226–9
pollution 192, 196, 200, 202
  plastic 192, 200, 248, 257
Porbeagle 89, 110–11, 122, 125, 130, 226
Poroderma africanum 225
Port Jackson Shark 23, 61, 68, 72, 76–7, 184
Portuguese Dogfish 154
predator avoidance 42, 59, 70, 138, 157, 190, 225, 244
pressure 154

prey/predation 23, 35, 54–7, 59, 70–1, 86–90, 106, 125, 214–15
  Basking Shark 46
  Blackmouth Catshark 172
  Bonnethead 204
  café 123
  Caribbean Reef Shark 234
  Cookiecutter Shark 164
  daily ration 90
  Goblin Shark 166
  Lemon Shark 221
  Nurse Shark 40
  Sand Tiger 236
  strange items 122–3
  see also apex predators; mesopredators
Prickly Shark 68
Prionace glauca see Blue Shark
Pristiophorus cirratus 82–3
Pristis pectinata see Smalltooth Sawfish
Puffadder Shyshark 224
punting 52
purse seines 255, 257, 272
Pyjama Shark 225

Q

Queen Conch 40

R

radio tags 98
ram ventilation 230
rays 14, 24, 36, 38, 112, 152, 179, 194, 211, 250, 268
rectal gland 179, 180, 181
reproduction 26, 28–9, 244–5
  adaptations 64–9
  Basking Shark 46
  birth 66–7, 82, 105, 130
  Blue Shark 138
  copulation 65
  deep-sea sharks 161, 163
  fecundity 28–9, 38, 68, 102, 105, 259

gestation periods 105, 168, 232
Great Hammerhead 44
internal fertilization 26, 51, 65
life histories 130
Little Gulper Shark 170
mating 114
multiple paternity 38
parthenogenesis 204
Sand Tiger 236
sexual maturity 102, 104–5
Swell Shark 42
Tiger Shark 232
Whitespotted Bamboo Shark 238
Whitetip Reef Shark 114
yolk-sac viviparity 68, 78, 264, 268
requiem sharks 67, 89, 211
Rhincodon typus see Whale Shark
Rhizoprionodon terraenovae see Atlantic Sharpnose Shark
ridgeback sharks 198
"roadkill" 248–9
rolling 50
rostrum 82, 112

S

Salmon Shark 122, 125, 130, 132, 136–7
Sand Tiger 67, 68, 105, 236–7
Sandbar Shark 72, 98, 101, 104, 184, 187–9, 190, 198–9, 256
satellite (SAT) tags 98, 99
sawsharks 66–7, 82
Scalloped Hammerhead 52, 214, 255, 260
School Shark 96
Scyliorhinus canicula see Small-spotted Catshark
sea-level rise 194–5
seagrass 208, 209, 210, 222, 260
seal attacks 70, 106, 225, 228

INDEX

sexual dimorphism 65
shagreen 46
shark bites/attacks 21, 134–5, 140, 176, 195, 196, 262–3
shark fin soup trade 198, 250–1
Shark Lab 215, 218–21
shark repellents 262
Shortfin Mako 54–7, 60–1, 74–5, 101, 118, 121, 122, 125, 130, 134
Sicklefin Lemon Shark 214, 215
Silky Shark 122, 129, 130, 134, 255, 272–3
Silvertip Shark 86, 93, 210, 214, 215
Sixgill Sawshark 54
sixgill sharks 154, 161
  see also Bluntnose Sixgill Shark
size 102, 152
  largest sharks 46
  longest sharks 38
  sexual dimorphism 65
  sexual maturity 104
  White Shark 106
  see also growth
skates 14, 24, 36, 52, 152, 211
skull 31
sleeper sharks 157, 160, 226, 228
slingshot feeding 166
Small-spotted Catshark 68
Smalltooth Sawfish 86, 112–13
smart position and temperature (SPOT) transmitting tag 99
smell, sense of 59
Smooth Hammerhead 255
smooth-hound sharks 23, 65, 190, 250, 252–3, 266–7
Somniosus microcephalus see Greenland Shark
Somniosus pacificus 226

speciation 28, 29, 153
Sphyrna lewini see Scalloped Hammerheads
Sphyrna mokarran see Great Hammerheads
Sphyrna tiburo see Bonnethead
Sphyrna zygaena see Smooth Hammerhead
Spinner Shark 89, 190
Spiny Dogfish 50, 105, 185, 190, 252–3, 264–5
spiny "sharks" *see* acanthodians
spiracles 52
squalomorphs 36
Squalus acanthias see Spiny Dogfish
Squalus cubensis see Cuban Dogfish
Squatina squatina 268–9
Stegostoma fasciatum 89
Swell Shark 42–3
swim bladders 31
swimming adaptations 35, 50

T

tag-and-release fishing 95–101
taste buds 60
Tawny Nurse Shark 214
teeth 27, 32, 38, 65, 76, 82, 160
telemetry 96, 98–9, 105
temperature 150, 153, 222, 226–9
  behavioral thermoregulation 80
  freshwater 181
  see also climate change
Thames, River 185
thermocline 121, 150, 153
thermodynamics, laws of 89, 91
thresher sharks 38, 68, 120, 121, 123, 130, 134, 142–3, 255

Tiger Shark 61, 73, 86, 93, 104, 120, 210, 214, 215, 222, 232–3, 256
Triaenodon obesus see Whitetip Reef Shark
Triakis semifasciata 50
trimethylamine N-oxide (TMAO) 154, 229
trophic cascade 93
trophic levels 23, 89–93, 105, 215, 234, 243
trophic migration 123

U

Unicorn Shark 34
urea 179, 181, 184

V

Velvet Belly Lanternshark 54, 78–9
vertebrae 32, 35
vertical migration 101, 108, 118–19, 121, 129, 153–4
vision 59–60, 150, 157

W

water pressure 148, 150, 154
Whale Shark 23, 68, 90, 105, 120, 123, 129, 130, 144–5, 204, 249
White Shark 22, 23, 29, 37, 68, 86, 106–7, 125, 130, 178, 195, 223
  daily ration 90
  eyes 60
  heart 60
  lifespan 102
  migration 106, 122–3
  oophagy 67, 106
  seal attacks 70, 106
  sexual maturity 104
  trophic level 89
Whitespotted Bamboo Shark 50, 238–9
Whitetip Reef Shark 68, 114–15, 210, 214

Winyah Bay 184, 188–91
wobbegongs 144, 214

Y

yawing 50
yolk-sac viviparity 68, 78, 264, 268

Z

Zebra Shark 89

ACKNOWLEDGMENTS

We owe eternal gratitude to more people than this space allows. What a pleasure it has been working with this book's publishing team. Who knew producing a book took a village? We thus thank with great enthusiasm and admiration Kate Shanahan, Richard Webb, Tom Broadbent, Wayne Blades, Sarah Skeate, Les Hunt, and the nameless others at UniPress Books who assembled our words into a book that makes us proud! And the photographers – seeing their breathtaking work the first time in the book's page proofs left us, true to form, breathless!

Oh no! We're out of our allocation of words and we have not thanked the most important people in our lives, our families and dear friends, personal and professional. And the sharks! And the irreplaceable planet on which we all live!

PICTURE CREDITS

Alamy 12 Ken Kiefer 2/ Alamy; 24 left Jeff Rotman/Nature Picture Library/Alamy; 24 right Brandon Cole Marine Photography/Alamy Stock Photo; 26 left Tony Wu/Naturepl.com; 26 right Richard Robinson/Nature Picture Library/Alamy; 34 Arco/I. Schulz/Imagebroker/Alamy; 35 anbusiello TW/Alamy Stock Photo; 36 The Natural History Museum/Alamy Stock Photo; 41 Hans Leijnse/NiS/Minden Pictures/Alamy; 45 Ken Keifer/Cultura Creative Ltd/Alamy Stock Photo; 63 Sean Scott/RooM the Agency/Alamy Stock Photo; 75 WaterFrame/Alamy Stock Photo; 103 Franco Banfi/Nature Picture Library/Alamy Stock Photo; 104 Hugh Gentry/REUTERS/Alamy; 118 Chronicle/Alamy Stock Photo; 132 Jeff Rotman/Alamy Stock Photo; 157 Nature Picture Library/Alamy Stock Photo; 158 NOAA/Alamy Stock Photo; 162 Andia/Alamy Stock Photo; 163 Brandon Cole Marine Photography/Alamy Stock Photo; 167 Marko Steffensen/Alamy Stock Photo; 169 Solvin Zankl/Nature PL/Alamy Stock Photo; 176 James Featherby/Alamy Stock Photo; 224 BIOSPHOTO/Alamy Stock Photo; 233 Scubazoo/Alamy Stock Photo; 237 Martin Strmiska/Alamy Stock Photo; 249 Duncan Murrell/Robert Harding/Alamy Stock Photo; 260 Jeff Rotman/Alamy Stock Photo; 273 Martin Strmiska/Alamy Stock Photo.

Annie Guttridge 214; 221; 222; 223; 246.

Blue Planet Archive 9 BluePlanetArchive/Lisa Collins; 29 BluePlanetArchive/Doug Perrine; 39 BluePlanetArchive/Eric Cheng; 43 BluePlanetArchive/Andy Murch; 50 BluePlanetArchive/imageBROKER/Norbert Probst; 52 left BluePlanetArchive/Marc Chamberlain; 54 BluePlanetArchive/C & M Fallows; 60 BluePlanetArchive/C & M Fallows; 71 BluePlanetArchive/Mark Conlin; 72-73 BluePlanetArchive/Jeff Rotman; 77 BluePlanetArchive/Doug Perrine; 79 BluePlanetArchive/Espen Rekdal; 81 BluePlanetArchive/Andy Murch; 83 BluePlanetArchive/Marty Snyderman; 107 BluePlanetArchive/C & M Fallows; 109 BluePlanetArchive/Bruce Rasner; 111 BluePlanetArchive/Doug Perrine; 120 BluePlanetArchive/Doug Perrine; 125 BluePlanetArchive/Doug Perrine; 130 BluePlanetArchive/Doug Perrine; 133 BluePlanetArchive/Jeff Rotman; 141 BluePlanetArchive/Andy Murch; 149 BluePlanetArchive/Andy Murch; 150 right BluePlanetArchive/Doug Perrine; 152 BluePlanetArchive/Gwen Lowe; 153 BluePlanetArchive/Makoto Hirose/e-Photo; 154 BluePlanetArchive/John Morrissey; 156 BluePlanetArchive/Kelvin Aitken/VWPics; 159 BluePlanetArchive/Doug Perrine; 160 BluePlanetArchive/Gwen Lowe; 161 BluePlanetArchive/Andy Murch; 165 BluePlanetArchive/Jeff Milisen; 173 Blue Planet Archive/Andy Murch; 180 BluePlanetArchive/Doug Perrine; 182 BluePlanetArchive/Andy Murch; 186 BluePlanetArchive/Doug Perrine; 199 BluePlanetArchive/Doug Perrine; 203 BluePlanetArchive/Doug Perrine; 205 BluePlanetArchive/Masa Ushioda; 220 BluePlanetArchive/Marilyn & Maris Kazmers; 239 BluePlanetArchive/Doug Perrine; 251 BluePlanetArchive/Makoto Hirose/e-Photo; 255 BluePlanetArchive/George McCallum; 267 & 269 BluePlanetArchive/Andy Murch; 271 BluePlanetArchive/Toshio Minami/e-Photo.

Bimini Biological Field Station 217.

CCU Shark Project 189.

CSULB Shark Lab 70.

Dean Grubbs 101.

Dr Kelly Kingon 105 Dr Kelly Kingon, The University of Trinidad and Tobago.

iStockphoto 27 s1murg/iStockphoto; 56 Alessandro De Maddalena/iStockphoto; 115 atese/iStockphoto; 126 Madelein Wolf/iStockphoto; 148 rusm/iStockphoto; 212 atese/iStockphoto.

Lesley Rochat 10; 32-33; 263.

Nature Picture Library 137 Andy Murch/Naturepl.com; 143 Richard Herrmann/Naturepl.com; 171 Gary Bell/Oceanwide/Naturepl.com; 197 Brandon Cole/Naturepl.com; 201 Andy Murch/Naturepl.com; 247 Ralph Pace/Naturepl.com; 265 Andy Murch/Naturepl.com.

NOAA Fisheries 98 NOAA Fisheries; 99 Matt Ellis/NOAA Fisheries.

Shane Gross 64; 65; 119; 150 left; 208; 209; 215; 216; 218; 226–227; 244; 245.

Shutterstock 3 Rich Carey/Shutterstock; 4 (from top to bottom) BearFotos/Shutterstock, Rich Carey/Shutterstock, Good luck images/Shutterstock, frantisekhojdysz/Shutterstock; 5 top frantisekhojdysz/Shutterstock; 5 bottom IrinaK/Shutterstock; 11 nicolasvoisin44/Shutterstock; 14 wildestanimal/Shutterstock; 15 top Kim_Briers/Shutterstock; 15 bottom Andrea Izzotti/Shutterstock; 16 top Mohamed AlQubaisi/Shutterstock; 16 bottom Susana_Martins/Shutterstock; 17 Alex Rush/Shutterstock; 20 Frolova_Elena/Shutterstock; 21 Martin Prochazkacz/Shutterstock; 28 Charlotte Bleijenberg/Shutterstock; 52 right frantisekhojdysz/Shutterstock; 61 Nathaneal Largent/Shutterstock; 62 Ciril Monteiro/Shutterstock; 68 Kolf/Shutterstock; 69 HollyHarry/Shutterstock; 102 vkilikov/Shutterstock; 103 Dotted Yeti/Shutterstock; 122 Rich Carey/Shutterstock; 124 Martin Prochazkacz/Shutterstock; 127 Jennifer Mellon Photos/Shutterstock; 128 martin_Hristov/Shutterstock; 145 Onusa Putapitak/Shutterstock; 179 Mr. Meijer/Shutterstock; 183 Leonardo Gonzalez/Shutterstock; 185 Joern_k/Shutterstock; 192 Junk Culture/Shutterstock; 194 iofoto/Shutterstock; 210 Kristina Vackova/Shutterstock; 231 Fiona Ayerst/Shutterstock; 235 Shane Gross/Shutterstock; 248 VisionDive/Shutterstock; 251 inset dangdumrong/Shutterstock; 253 top Rafeed Hussain/Shutterstock; 253 bottom spatuletail/Shutterstock; 254 Tara Lambourne/Shutterstock; 258 Ethan Daniels/Shutterstock; 262 Erin Bouda/Shutterstock.

Tanya Houppermans 2; 30; 47; 53; 58; 86; 88; 90; 91; 92; 94; 95; 97; 139; 184; 242; 274-275; 283.

All reasonable efforts have been made to trace copyright holders and to obtain their permission for the use of copyright material. The publisher apologizes for any errors or omissions in the list above and will gratefully incorporate any corrections in future reprints if notified.